Office of the Deputy Prime Minister

Creating sustainable communities

D0550640

Environmental Impact Assessment

A guide to procedures

November 2000

Thomas Telford: London

Following the reorganisation of the government in May 2002, the responsibilities of the former Department of the Environment, Transport and the Regions (DETR) and latterly Department for Transport, Local Government and the Regions (DTLR) in this area were transferred to the Office of the Deputy Prime Minister.

Office of the Deputy Prime Minister
Eland House
Bressenden Place
London SWIE 5DU
Telephone 020 7944 4400
Web site www.odpm.gov.uk/

Further copies of this report are available from:

T Thomas Telford *Publishing*
The Customer Services Department
Thomas Telford Limited, Units I/K
Paddock Wood Distribution Centre
Paddock Wood, Tonbridge
Kent, TN12 6UU
Tel: 01892 832299
Fax: 01892 837272

http://www.thomastelford.com

ISBN 0 7277 2960 8

Printed in Great Britain on material containing 75% post-consumer waste and 25% ECF pulp
November 2000
Amended reprint 2004

CONTENTS

Introduction

Environmental impact assessment (EIA) is an important procedure for ensuring that the likely effects of new development on the environment are fully understood and taken into account before the development is allowed to go ahead. This booklet, which is intended primarily for developers and their advisers, explains how European Community (EC) requirements for the environmental impact assessment of major projects have been incorporated into consent procedures in the UK. It revises the booklet 'Environmental Assessment: A Guide to the Procedures', first published in 1989, to take account of the requirements of Directive 97/11/EC, which was adopted on 3 March 1997 and came into effect on 14 March 1999.

Directive 97/11/EC amends the original Directive 85/337/EEC on 'The assessment of the effects of certain public and private projects on the environment', which came into effect in July 1988. The text of the Directive as amended (with amendments shown in italics) is reproduced in Appendix 1 to this booklet. Throughout this booklet, references to 'the Directive' mean Directive 85/337/EEC as amended by Directive 97/11/EC. The effect of the Directive is to require environmental impact assessment to be carried out, before development consent is granted, for certain types of major project which are judged likely to have significant environmental effects.

Parts 1 and 2 of this booklet explain the procedures which apply to projects falling within the scope of the Directive and requiring planning permission in England and Wales. For such projects the Directive was given legal effect through the Town and Country Planning (Environmental Impact Assessment) (England and Wales) Regulations 1999 (SI No. 293) ('the EIA Regulations') which came into force on 14 March 1999 and apply to relevant planning applications lodged on or after that date. The full text of the Regulations is published separately by The Stationery Office. For ease

5

of reference, those parts of the Regulations which list the types of project to which they apply, and specify what information an environmental statement must contain, are reproduced in Appendices 2 to 4 to this booklet. The booklet is not intended to be an authoritative interpretation of the law and does not remove the need to refer to the Regulations.

Formal guidance on procedures under the Regulations, directed principally at local planning authorities, was issued in DETR Circular 2/99 (Welsh Office Circular 11/99). Although the present booklet, like its predecessor, is meant to be reasonably self-contained, developers may need to refer to that Circular, particularly for fuller information on how planning authorities are expected to judge the significance of a project's likely effects for the purpose of deciding whether environmental impact assessment is required.

Parts 1 and 2 also give some general guidance, applicable to all types of project, on the nature of environmental impact assessment and on the practical aspects of preparing an environmental statement.

Part 3 gives a brief account of the procedures which apply to other projects within the scope of the Directive but which are not approved under planning procedures, for example, motorways, harbour works and long distance pipe-lines. It also deals very briefly with environmental impact assessment procedures in Scotland and Northern Ireland. For the detailed requirements, reference will need to be made to the relevant statutory instruments and associated guidance (see Appendix 8 to this booklet).

Throughout this booklet, the term 'environmental impact assessment' (EIA) is used to describe the whole process whereby information about the environmental effects of a project is collected, assessed and taken into account in reaching a decision on whether the project should go ahead or not. The process was formerly referred to in the UK as 'environmental assessment' (EA). An 'environmental statement' is a document setting out the developer's own assessment of a project's likely environmental effects, which is prepared and submitted by the developer in conjunction with the application for consent.

1 Environmental impact assessment and projects which require planning permission

What is environmental impact assessment?

1   The term 'environmental impact assessment' (EIA) describes a procedure that must be followed for certain types of project before they can be given 'development consent'. The procedure is a means of drawing together, in a systematic way, an assessment of a project's likely significant environmental effects. This helps to ensure that the importance of the predicted effects, and the scope for reducing them, are properly understood by the public and the relevant competent authority before it makes its decision.

2   Environmental impact assessment enables environmental factors to be given due weight, along with economic or social factors, when planning applications are being considered. It helps to promote a sustainable pattern of physical development and land and property use in cities, towns and the countryside. If properly carried out, it benefits all those involved in the planning process.

3   From the developer's point of view, the preparation of an environmental statement in parallel with project design provides a useful framework within which environmental considerations and design development can interact. Environmental analysis may indicate ways in which the project can be modified to avoid possible adverse effects, for example through considering more environmentally-friendly alternatives. Taking these steps is likely to make the formal planning approval stages run more smoothly.

4   For the planning authority and other public bodies with environmental responsibilities, environmental impact assessment

provides a basis for better decision making. More thorough analysis of the implications of a new project before a planning application is made, and the provision of more comprehensive information with the application, should enable authorities to make swifter decisions. While the responsibility for compiling the environmental statement rests with the developer, it is expected that the developer will consult those with relevant information, and the Regulations specifically require that public authorities which have information in their possession which is relevant to the preparation of the environmental statement should make it available to the developer.

5   The general public's interest in a major project is often expressed as concern about the possibility of unknown or unforeseen effects. By providing a full analysis of a project's effects, an environmental statement can help to allay fears created by lack of information. At the same time, early engagement with the public when plans are still fluid can enable developers to make adjustments which will help to secure a smoother passage for the proposed development and result in a better environmental outcome. The environmental statement can also help to inform the public on the substantive issues which the local planning authority will have to consider in reaching a decision. It is a requirement of the Regulations that the environmental statement must include a description of the project and its likely effects together with a summary in non-technical language. One of the aims of a good environmental statement should be to enable readers to understand for themselves how its conclusions have been reached, and to form their own judgements on the significance of the environmental issues raised by the project.

6   Environmental impact assessment can therefore be helpful to all those concerned with major projects. The following paragraphs describe the procedures for deciding whether EIA is necessary in a particular case and, where it is, for carrying out the assessment. The procedure is intended to make the most of the potential benefits of EIA, while keeping the process as simple and flexible as possible, and avoiding any duplication of existing planning procedures.

When is environmental impact assessment needed?

7   The Regulations apply to two separate lists of projects:

(i) 'Schedule 1 projects', for which EIA is required in every case;

(ii) 'Schedule 2 projects', for which EIA is required only if the particular project in question is judged likely to give rise to significant environmental effects.

Lists of Schedule 1 and Schedule 2 projects are given in, respectively, Appendices 2 and 3 to this booklet.

8   For Schedule 1 projects, whether or not a particular project falls within the scope of the Regulations will normally be clear: several of the definitions of Schedule 1 projects incorporate an indication of scale, in the form of a quantified threshold, which clearly identifies the projects requiring EIA. Where there is any doubt about a project's inclusion in Schedule 1, the procedures described in paragraphs 14–19 below can be used to obtain an opinion from the planning authority or a direction from the Secretary of State (or, in Wales, the National Assembly for Wales).

9   For the much longer list of Schedule 2 projects, the issue turns on the likelihood of 'significant environmental effects'. For the different types of project described in column 1 of Schedule 2, the 1999 Regulations introduced a system of thresholds and criteria, shown in column 2, as a method of discounting development which is not likely to have significant effects on the environment. For development where the applicable threshold or criterion is not exceeded or met, EIA is not normally required. However, even where the threshold or criterion is not met or exceeded, EIA may be required if the proposed development is in, or partly in, a 'sensitive area' (see paragraph 10). In exceptional circumstances the Secretary of State (or, in Wales, the National Assembly for Wales) may exercise his power under the Regulations to direct that a particular type of Schedule 2 development requires EIA even if it is not to be located in a sensitive area and does not exceed or meet the applicable threshold or criterion.

10   The more environmentally sensitive the location, the more likely it is that the effects of development will be significant and that EIA will be required. That is why the thresholds and criteria do not apply where development is proposed in, or partly in, a 'sensitive area' as

defined in the Regulations. Such areas include Sites of Special Scientific Interest (SSSIs), National Parks, Areas of Outstanding Natural Beauty, the Broads, World Heritage Sites and scheduled monuments. There is no general presumption that every Schedule 2 development in a sensitive area will require EIA. Nevertheless, in the case of development to be located in or close to SSSIs, especially those which are also international conservation sites such as Ramsar sites or Special Protection Areas for birds, the likely environmental effects will often be such as to require EIA.

How 'significance' will be assessed

11   Developments which meet or exceed the applicable threshold are considered on a case-by-case basis. For the purpose of determining whether EIA is necessary, those of the selection criteria set out in Schedule 3 to the Regulations which are relevant to the proposed development, must be taken into account. The selection criteria fall into the three broad headings: characteristics of the development, location of the development, and characteristics of the potential impact.

12   For obvious reasons there can be no general definition of what constitutes significance. General guidance on how to assess 'significance' is contained in DETR Circular 2/99 (Welsh Office Circular 11/99); and rulings may be obtained from the local planning authority or the Secretary of State (or, in Wales, the National Assembly for Wales) on whether EIA is required in particular cases. Essentially the Circular suggests that there are three main criteria of significance:

(i)   major developments which are of more than local importance;

(ii)   developments which are proposed for particularly environmentally sensitive or vulnerable locations;

(iii) developments with unusually complex and potentially hazardous environmental effects.

13   These are very general guidelines and, to assist in their application to particular cases, the Circular also sets out indicative thresholds and

criteria by reference to particular categories of development listed in Schedule 2 to the Regulations. These are reproduced in the last column of the Table in Appendix 3 to this booklet. It will be obvious that none of these guidelines can be applied as hard and fast rules; circumstances are bound to vary greatly from case to case. Some large-scale projects which exceed the indicative thresholds may not be significant enough to require EIA; some smaller projects, particularly in sensitive locations, may be candidates for EIA. Nevertheless, the guidance in the Circular should provide a starting point for consideration by the developer and the planning authority of the need for EIA. If the matter is referred to the Secretary of State (or, in Wales, the National Assembly for Wales), he will have regard to the published criteria.

Obtaining a ruling on the need for EIA

14   Where there is a possibility that a proposed development will require environmental impact assessment, developers are advised to consult the relevant planning authority well in advance of a planning application. Developers can decide for themselves that a given project falls within the scope of the Regulations so that an environmental statement will be needed. But the Regulations also provide a procedure which enables developers to apply to the planning authority for an opinion ('screening opinion') on whether EIA is needed in a particular case, as soon as a basic minimum of information can be provided about the proposal. This must include a plan on which the site of the proposed development is identified, and a brief description of its nature and purpose and of its possible effects on the environment. This can, of course, be supplemented with other information if the developer wishes.

15   Where such information can be provided, the developer may approach the planning authority at any time for an opinion on the need for EIA. This can be done well in advance of any formal planning application, although any approach to the planning authority before the planning application stage is entirely voluntary. Where such an approach is made, the planning authority must give its opinion within three weeks, unless the developer agrees to a longer period. The planning authority may request further information from the developer, but this in itself does not extend the three-week time

limit, unless the developer agrees. The planning authority must make its determination available for public inspection at the place where the appropriate register (or relevant section of that register) is kept.

16 Where the planning authority expresss the opinion that a particular proposal requires EIA – whether in response to a request from a developer prior to a planning application, or following a planning application – it must provide a written statement giving clear and precise reasons for its opinion. Both that statement and the developer's application for an opinion are then made available for public inspection at the same place as the register.

17 A developer who is dissatisfied with the planning authority's opinion that EIA is required may refer the matter to the Secretary of State (or, in Wales, the National Assembly for Wales). The developer is simply required to copy the relevant papers to the Government Office in the region concerned (or the Assembly) and add any representations which are considered to be appropriate in the light of the planning authority's statement. The Secretary of State (or the Assembly) will then normally give a direction within three weeks of the developer's application; and, if the direction is to the effect that EIA is required, it will be accompanied by a statement of reasons. The developer may, when requesting a screening opinion from the planning authority, simultaneously request an opinion on what should be included in any environmental statement (see paragraph 25 below).

18 The broad intention of this procedure is to ensure that developers can obtain a clear ruling on the need for EIA well before they reach the stage of lodging a formal planning application. This should minimise the possibility of delay or uncertainty at that stage. Where the matter is not raised until a formal planning application is lodged, the developer risks serious delay if either the planning authority or the Secretary of State (or, in Wales, the National Assembly for Wales) rules that an environmental statement must be prepared. No action will be taken on the planning application until the developer has prepared an environmental statement and submitted it to the planning authority.

19 In most cases this procedure will give developers a firm decision on the need for EIA as soon as they can provide basic information about their project. There may occasionally be cases where the receipt of information directly leads the Secretary of State (or, in Wales, the National Assembly for Wales) to issue a direction when there has been no request for one from the developer; or to overrule a planning authority's opinion that EIA is not required; or even, very exceptionally, to reverse an earlier direction. This could, for instance, happen where a development proposal came to the Secretary of State's (or the Assembly's) notice on 'call-in'; or where third party representations drew attention to aspects of a proposed development which were not known to the Secretary of State (or the Assembly) when the initial direction was given. Exceptionally, also, a planning authority which has given a pre-application opinion that EIA is not required might consider it necessary to reverse that decision when the planning application is formally submitted. In that case the developer could, of course, apply to the Secretary of State (or the Assembly) for a direction.

'Permitted development rights' (PDRs)

20 Developments which do not require planning permission because of the provisions of the Town and Country Planning (General Permitted Development) Order 1995 (SI No. 418) continue to enjoy PDRs provided that they do not fall into Schedule 1 or 2 of the EIA Regulations. For developments that do fall within Schedule 1 or 2, the general position is as follows. Schedule 1 projects are not permitted development, and always require the submission of a planning application and an environmental statement. PDRs for Schedule 2 projects which either exceed or meet the applicable threshold or criterion, or are wholly or partly in a sensitive area, are also withdrawn, unless the local planning authority has adopted a screening opinion (or the Secretary of State (or, in Wales, the National Assembly for Wales) has directed) to the effect that EIA is not required. There are exceptions to these provisions in the case of the following classes in Schedule 2 to the 1995 Order: Part 7, Class D of Part 8, Part 11, Class B of Part 12, Class F(a) of Part 17, Class A of Part 20, Class B of Part 20, and Class B of Part 21. These exceptions exist for a variety of reasons. For example, some relate to projects subject to alternative consent procedures, and others to projects begun before Directive 85/337/EEC came into operation.

Further information about the exceptions, and about permitted development and EIA generally, can be found in DETR Circular 2/99 (Welsh Office Circular 11/99) paragraphs 61–65 and 151–156.

Simplified planning zones and enterprise zones

21 Special considerations apply to projects proposed for simplified planning zones (SPZs) and enterprise zones (EZs).

Simplified planning zones

22 All Schedule 1 projects are excluded from the scope of SPZs and therefore require EIA as part of the application for planning permission. Where the terms of SPZ schemes would permit Schedule 2 projects to be undertaken without specific planning permission, developers must notify the planning authority that they intend to undertake such development. This notification will give the planning authority and, where appropriate, the Secretary of State (or, in Wales, the National Assembly for Wales) the opportunity to consider whether the project is likely to give rise to significant environmental effects. If so, planning permission will be required in the normal way and an environmental statement must be prepared. If not, the project will enjoy the benefit of the general permission granted by the SPZ scheme, and no separate application for planning permission will be necessary.

Enterprise zones

23 If the planning scheme allows for Schedule 1 projects, EIA must be carried out. As regards Schedule 2 projects, the same procedures apply as those for SPZs described above.

24 The arrangements in SPZs and EZs are explained in DETR Circular 2/99 (Welsh Office Circular 11/99).

2 Preparing an environmental statement: the planning procedures

25   The developer is responsible for preparing the environmental statement which must be submitted with the planning application. A developer may choose to engage consultants for some or all of the work. The preparation of the statement should be a collaborative exercise involving discussions with the local planning authority, statutory consultees and possibly other bodies as well. There is no prescribed form of statement, provided that the requirements of the Regulations are met. The Regulations enable a developer, before making a planning application, to ask the local planning authority for its formal opinion ('scoping opinion') on the information to be included in an environmental statement. The developer can therefore be clear as to what the authority considers are the aspects of the environment which would be affected. The request may be made at the same time as for the screening opinion, and must be accompanied by the same information as for the screening opinion (paragraph 14 above). A developer may also wish to submit a draft outline of the environmental statement, indicating what the developer thinks the main issues are, as a focus for the authority's consideration. The authority must consult certain bodies (see paragraph 36) and the developer before adopting a scoping opinion, and must adopt the scoping opinion within five weeks of receiving the request. This period may be extended provided that the developer agrees. The scoping opinion must be kept available for public inspection for two years, with the request (including documents submitted by the developer as part of that request), at the place where the planning register is kept.

26   There is no provision for an appeal to the Secretary of State (or, in Wales, the National Assembly for Wales) if the developer and local planning authority disagree about the content of an environmental

statement. If an authority fails to adopt a scoping opinion within five weeks (or any agreed extension), the developer may apply to the Secretary of State (or the Assembly) for a scoping direction. The application must be accompanied by the previous documents relating to the request for a scoping opinion, plus any additional representations the developer wishes to make. The developer should also send a copy of the request and any representations to the authority. The Secretary of State (or the Assembly) must make a scoping direction within five weeks of receiving the request, or such longer period as he may reasonably require. He must consult the developer and certain bodies beforehand. A copy of the scoping direction will be sent to the developer and the authority.

27   The aim should be to provide as systematic and objective an account as is possible of the significant environmental effects to which the project is likely to give rise. Where the statement embodies or summarises the conclusions of more detailed work, sufficient information should be provided to enable those who wish to do so to verify the statement's conclusions and to identify the source of the information provided. The environmental statement must contain a non-technical summary which will enable non-experts to understand its findings.

Preliminary consultations

28   One of the main emphases of the process of environmental impact assessment is on the need for fuller and earlier consultation by the developer with bodies which have an interest in the likely environmental effects of the development proposal. If important issues are not considered at a very early stage, they may well emerge when a project's design is well advanced, and necessitate rethinking and delay. Ideally, EIA should start at the stage of site selection and (where relevant) process selection, so that the environmental merits of practicable alternatives can be properly considered.

29   While a developer is under no formal obligation to consult about the proposal before the submission of a formal planning application, there are good practical reasons for doing so. Authorities will often possess useful local and specialised information which is relevant to a project's

design, and officers may be able to give preliminary advice about local problems and about those aspects of the proposal that are likely to be of particular concern to the authority.

30 The timing of such informal consultations is at the developer's discretion; but it will generally be advantageous for them to take place as soon as the developer is in a position to provide sufficient information about the proposal to form a basis for discussion. The developer can ask that any information provided at this preliminary stage should be treated in confidence by the planning authority and any other consultees. If the developer is seeking a formal opinion from the planning authority on the need for environmental impact assessment (see paragraph 14), the information about the project which accompanies that request will be made public by the authority.

Content of the environmental statement

31 Developers and authorities should discuss the scope of an environmental statement before its preparation is begun. The formal requirements as to the content of environmental statements are set out in Schedule 4 to the Regulations, which is reproduced in Appendix 4 to this booklet. The information is given under separate headings: Part I and Part II. The statement must include at least the information included in Part II, and such of the information in Part I as is reasonably required to assess the environmental effects of the development and which the applicant can reasonably be required to complete. As a practical guide to the range of issues which may need to be considered, developers may find it helpful to use the checklist at Appendix 5 to this booklet as a basis for their discussions with the planning authority. The checklist is not meant to be regarded as a prescribed framework for all environmental statements. Its main purpose is to act as a guide or agenda for the preliminary discussions about the scope of the statement.

32 The comprehensive nature of the checklist at Appendix 5 should not be taken to imply that all environmental statements should cover every conceivable aspect of a project's potential environmental effects at the same level of detail. They should be tailored to the nature of the project and its likely effects. Whilst every environmental

statement should provide a full factual description of the project, the emphasis of Schedule 4 is on the main or significant effects to which a project is likely to give rise. In some cases, only a few of the aspects set out in the checklist will be significant in this sense and will need to be discussed in the statement in any great depth. Other issues may be of little or no significance for the particular project in question, and will need only very brief treatment, to indicate that their possible relevance has been considered.

33   It should be noted that developers are now required to include in the environmental statement an outline of the main alternative approaches to the proposed development that they may have considered, and the main reasons for their choice. It is widely regarded as good practice to consider alternatives, as it results in a more robust application for planning permission. Also, the nature of certain developments and their location may make the consideration of alternatives a material consideration. Where alternatives are considered, the main ones must be outlined in the environmental statement.

34   Even where a local planning authority has adopted a scoping opinion, the developer is responsible for the content of the statement which is finally submitted. Developers should bear in mind that planning authorities have powers to call for additional information when considering environmental statements and planning applications, and that they are likely to use those powers if they consider that aspects of a submitted environmental statement are inadequate (see paragraph 48 below). There is no provision for any disagreement between the developer and the planning authority over the content of an environmental statement to be referred to the Secretary of State (or, in Wales, the National Assembly for Wales), except through normal planning appeal procedures (see paragraph 52 below).

35   Developers should consider at an early stage whether an assessment of environmental effects may also be required under another European Community Directive, such as the Habitats Directive (92/43/EEC), the Wild Birds Directive (79/409/EEC), the Integrated Pollution Prevention and Control Directive (96/61/EC) or the Control of Major Accident Hazards Directive (96/82/EC). Although the requirements of these and of the EIA Directive are all independent of each other,

there are clearly links between them. Where more than one regime applies, developers could save unnecessary time and effort if they identify and co-ordinate the different assessments required. Advice on the links between the EIA system and the requirements of the Habitats Regulations is offered in PPG 9 on Nature Conservation (or, in Wales, Planning Guidance (Wales) Planning Policy First Revision), and on the links between the Town and Country Planning system and the IPPC authorisation system in PPS 23 on Planning and Pollution Control (or, in Wales, Planning Guidance (Wales) Planning Policy First Revision and Planning Guidance (Wales) Technical Advice Note (Wales) 5 'Nature Conservation and Planning').

Statutory and other consultees; the general public

36   The Regulations give a particular role in environmental impact assessment to those public bodies with statutory environmental responsibilities who must be consulted by the planning authority before a Schedule 1 or a Schedule 2 planning application is determined. A full list of these statutory consultation bodies is given in Appendix 6 to this booklet.

37   Where the planning authority or the Secretary of State (or, in Wales, the National Assembly for Wales) rules that EIA is required, those bodies which are statutory consultees for the particular project in question will be notified and the developer will be informed accordingly. The effect of this notification is to put those bodies under an obligation to provide the developer (on request) with any information in their possession which is likely to be relevant to the preparation of the environmental statement. An example might be information held by English Nature (or, in Wales, the Countryside Council for Wales) about the ecology of a particular area, which could be relevant to the assessment of a project's likely effects.

38   It is up to the developer to approach the statutory consultees and indicate what sort of information would be helpful in preparing the environmental statement. The obligation on statutory consultees relates only to information already in their possession; they are not required to undertake research on behalf of the developer. Nor, at this stage, would consultees be expected to express a view about the

merits of the proposal; their views on merits are invited at a later stage (see paragraph 46 below). Consultees may make a reasonable charge to cover the cost of making information requested by a developer available.

39 Developers should also consider whether to consult the general public, and non-statutory bodies concerned with environmental issues, during the preparation of the environmental statement. Bodies of this kind may have particular knowledge and expertise to offer. Some are national organisations, for instance, the Royal Society for the Protection of Birds; in most areas there are also active local amenity societies and environmental groups. While developers are under no obligation to publicise their proposals before submitting a planning application, consultation with local amenity groups and with the general public can be useful in identifying key environmental issues, and may put the developer in a better position to modify the project in ways which would mitigate adverse effects and recognise local environmental concerns. It will also give the developer an early indication of the issues which are likely to be important at the formal application stage if, for instance, the proposal goes to public inquiry.

Techniques of assessment; sources of advice

40 Extensive literature is available on how to assess the effects on the environment of particular processes and activities. The assessment techniques used, and the degree of detail in which any particular subject is treated in an environmental statement, will depend on the character of the proposal, the environment which it is likely to affect, and the information available. While a careful study of the proposed location will generally be needed (including environmental survey information), original scientific research will not normally be necessary. The local planning authority and statutory consultees may be able to advise the developer on sources of specialist information, for example, about particular local conditions.

41 Environmental statements will often need to recognise that there is some uncertainty attached to the prediction of environmental effects. Where there is uncertainty, it needs to be explicitly recognised.

Uncertainty is not in itself a reason for discounting the importance of particular potential environmental effects, simply because other effects can be more confidently predicted.

Submission of environmental statement with planning application

42   To enable a planning application to be processed as quickly as possible, it is in the developer's interest to submit an environmental statement at the same time as the application is made. It will be for the planning authority to judge how much information is required in the particular case, but the preparation of an environmental statement is bound to require the developer to work out proposals in some detail; otherwise any thorough appraisal of likely effects will be impossible. Where an application is in outline, the planning authority will still need to have sufficient information on a project's likely effects to enable it to judge whether the development should take place or not. The information given in the environmental statement will have an important bearing on whether matters may be reserved in an outline permission; it will be important to ensure that the development does not take place in a form which would lead to significantly different effects from those considered at the planning application stage.

43   When the developer submits an environmental statement at the same time as the the planning application, three further copies must also be submitted for onward transmission by the planning authority to the Secretary of State (or, in Wales, the National Assembly for Wales). The developer is also required to provide the planning authority with sufficient copies of the environmental statement to enable one to be sent to each of the statutory consultees. Alternatively, the developer may send copies of the statement directly to the consultees. When submitting the application, the developer must inform the planning authority of the name of every body – whether or not it is a statutory consultee – to which a copy of the statement has been sent.

44   The developer should make a reasonable number of copies of the statement available for members of the public. A reasonable charge reflecting printing and distribution costs may be made.

Handling by the planning authority

45 Where the environmental statement is submitted with the planning application, the local planning authority will arrange for a notice to be published in a local newspaper and displayed at or near the site of the proposed development. If the environmental statement is submitted after the planning application, responsibility for publicising it falls to the developer.

46 The planning authority will place the planning application on Part 1 of the planning register, together with the environmental statement. The authority and the developer may wish to consider the need for further publicity at this stage, for example, publication of further details of the project in a local newspaper, or an exhibition. The planning authority will also need to notify statutory consultees of the application (unless the developer has already done so) and invite them to comment on the environmental statement. Consultees must be allowed at least 14 days from receipt of the statement in which to comment before a decision is taken. It will often be useful for the planning authority to discuss the project with consultees who have a particular interest in its environmental effects before reaching its conclusions on the planning applications.

47 The copies of the environmental statement that are forwarded by the planning authority to the Secretary of State (or, in Wales, the National Assembly for Wales), will assist in monitoring. In those exceptional circumstances where a proposed development is likely to have significant effects on the environment in another country, they will also enable certain international obligations for exchange of information with other countries to be met.

Requests for further information

48 Where the planning authority considers that the information provided in the developer's environmental statement, together with that available to the authority from other sources, is insufficient to permit a proper evaluation of the project's likely environmental effects, the authority can require further information, or evidence to verify the information that has already been provided. The use of these powers should not normally be necessary, especially if the parties have

worked together during the preparation of the environmental statement. Nevertheless, further consultation between the planning authority and the developer may be necessary at this stage, in particular to consider comments made by consultees and, possibly, amendments to the proposal to meet objections that have been raised.

49 Where an authority considers that it does not have the necessary expertise to evaluate the information contained in an environmental statement, it may decide to seek advice from consultants or other suitably qualified persons or organisations.

Determination of application

50 The planning authority is required to determine a planning application which is the subject of environmental impact assessment within 16 weeks from the date of receipt of the environmental statement, unless the developer agrees to a longer period. In determining the application, the authority is, of course, required to have regard to the environmental statement, as well as to other material considerations. As with any other planning application, the planning authority may refuse permission or grant it with or without conditions.

51 The planning authority cannot take the view that a planning application is invalid because it considers that an inadequate environmental statement has been submitted or because the developer has failed to provide any further information required under the powers described in paragraph 48 above. However, if the developer fails to provide enough information to complete the environmental statement, the application can be determined only by refusal.

Appeals and call-ins

52 The right of appeal to the Secretary of State (or, in Wales, the National Assembly for Wales) against an adverse decision by a planning authority (or against an authority's failure to determine an application within the 16-week time limit) is the same for planning

applications to which the EIA Regulations apply as for other applications. Similarly, the Secretary of State's (or the Assembly's) power to call in a planning application applies in these cases. Where an environmental statement has been prepared to accompany a planning application, the information which it contains will be among the material considerations which an Inspector will take into account in considering an appeal. The Secretary of State (or the Assembly) and Inspectors, like the planning authority, have power to request the developer to provide further information where they consider that the environmental statement is inadequate as it stands. Any additional information provided by the developer in response to such a request will be made available to all parties to an appeal.

Procedural stages

53 Appendix 7 to this booklet provides illustrative flow charts for the five main procedural stages: application to the planning authority for a screening opinion; application to the Secretary of State (or, in Wales, the National Assembly for Wales) for a screening direction; application to the planning authority for a scoping opinion; application to the Secretary of State (or the Assembly) for a scoping direction; and submission of an environmental statement to the planning authority in conjunction with a planning application.

3 Arrangements for other projects

54 The advice given in Parts 1 and 2 of this booklet is generally applicable to all projects. However, those Parts specifically describe the arrangements for EIA of projects in England and Wales which are approved through the planning system. Part 3 gives brief guidance on the legislative provisions on EIA for projects insofar as they are outside the scope of the planning system.

55 For all the following projects the Regulations described, which implement the Directive, include provisions for seeking a scoping opinion from the competent authority, for consulting statutory and other bodies, for the submission of an environmental statement in the prescribed form, and for publicising the statement. Where applicable, the Regulations disapply any thresholds for Annex II projects which are at least partly in a 'sensitive area', this term being defined in the same or almost the same way as in the planning EIA Regulations (see paragraph 10 above).

Projects not subject to planning control

The trunk road network

56 EIA is mandatory for projects to construct new motorways and certain other roads, including those with four or more lanes, and for certain road improvements (see Directive Annex I.7(b) and (c)). Projects to construct roads other than those in Annex I automatically fall within Annex II.10(e).

57 Roads in England and Wales for which the Secretary of State for Transport or the National Assembly for Wales is the highway authority, are approved under procedures set out in the Highways Act 1980. These roads – the trunk road network – currently include nearly all motorways. The Department concerned will normally

consult the public widely about alternative routes before selecting a preferred route for a new road. Once the preferred route has been announced, detailed design work is carried out leading to the publication of statutory orders. In most cases these orders are the subject of a public inquiry held by an independent inspector.

58   For these roads, the Directive has been implemented in England and Wales by the Highways (Assessment of Environmental Effects) Regulations 1999 (SI No. 369). These Regulations replace the Highways (Assessment of Environmental Effects) Regulations 1988 (SI No. 1241), except for projects begun before the new Regulations came into force on 13 March 1999.

59   The Regulations incorporate an exclusive threshold for the purpose of screening Annex II projects: only those in which the area of the completed works and the area occupied during the road construction or improvement period exceeds one hectare, or are wholly or partly in a sensitive area, need to be screened for EIA.

60   Roads developed by local planning authorities, and roads developed by private developers, require planning permission and the provisions described in Parts 1 and 2 of this booklet apply to such roads.

Oil and gas pipe-lines

61   Oil and gas pipe-lines with a diameter of more than 800 millimetres and a length of more than 40 kilometres fall within Annex I of the Directive. Pipe-line installations which are below either of these thresholds fall within Annex II. Oil and gas pipe-lines 10 miles long or less are approved under planning legislation.

Onshore pipe-line works

62   Proposed onshore pipes (but not those of public gas transporters – see below) which are more than 10 miles long require a pipe-line construction authorisation (PCA) from the Secretary of State for Trade and Industry under the Pipe-lines Act 1962. The Pipe-line Works (Environmental Impact Assessment) Regulations 2000 (SI No. 1928), which replace the provisions relating to pipe-lines in the Electricity and Pipe-line Works (Assessment of Environmental Effects) Regulations 1990, implement the Directive in England,

Wales and Scotland in respect of pipe-lines which require a PCA. They require 'relevant pipe-line works' to be subject to EIA unless the Secretary of State makes a determination that EIA is not necessary.

Public gas transporter pipe-line works

63   Under the licensing regime introduced by the Gas Act 1995, companies wishing to convey gas may be licensed as public gas transporters. As such, they are exempt from the need to obtain authorisation for their pipes under the Pipe-lines Act 1962.

64   The Public Gas Transporter Pipe-line Works (Environmental Impact Assessment) Regulations 1999 (SI No. 1672) came into force on 15 July 1999 and implement the Directive in England, Wales and Scotland. They require a public gas transporter proposing to undertake pipe-line works which fall in Annex I of the Directive to submit an environmental statement and apply to the Secretary of State for Trade and Industry for consent to carry them out. Pipe-line works in Annex II of the Directive may be subject to EIA if they have a design operating pressure exceeding 7 bar gauge or either wholly or in part cross a sensitive area. In these circumstances, the public gas transporter must, before commencing construction, either obtain a determination from the Secretary of State that an environmental statement is not required, or give notice that it intends in any event to produce an environmental statement. The Regulations also provide for the Secretary of State to require an environmental statement where proposed works do not meet these criteria but nevertheless it is considered that there are likely to be significant environmental effects.

Offshore oil and gas projects

65   Annex I.14 of the Directive covers the extraction of petroleum and natural gas for commercial purposes where the amount extracted exceeds 500 tonnes per day in the case of petroleum and 500 000 cubic metres per day in the case of gas.

66   For such projects, the Offshore Petroleum Production and Pipe-lines (Assessment of Environmental Effects) Regulations 1999 (SI No. 360) implement the Directive. They apply to the whole of the UK. They

replace the Offshore Petroleum and Pipe-lines (Assessment of Environmental Effects) Regulations 1998 (SI No. 968), which are revoked except for saving provisions for applications for consent received before 14 March 1999.

67 The Regulations are linked to licences granted by the Secretary of State for Trade and Industry under the Petroleum Act 1998. Such licences require the Secretary of State's consent to be obtained to the drilling of a well, the getting of petroleum (where the amount exceeds 500 tonnes per day in the case of oil and 500 000 cubic metres per day in the case of gas) and the erection of any structure in connection with a development. They require, subject to an exception for floating installations whose use commenced before 30 April 1998 (when the 1998 Regulations came into force), the Secretary of State's consent for the use of a floating installation in prescribed circumstances. The Regulations also require the Secretary of State's consent to the use of a mobile installation for the purposes of the extraction of petroleum where the main purpose of such extraction is the testing of a well.

68 For projects which fall outside Annex I of the Directive, the Secretary of State decides whether the project is likely to have significant effects and therefore whether EIA is required.

Nuclear power stations

69 Annex I.2 of the Directive covers nuclear power stations and other nuclear reactors, and now includes their dismantling or decommissioning, so EIA is mandatory for these projects. It does not cover research installations for the production and conversion of fissionable and fertile material, whose maximum power does not exceed 1 kilowatt continuous thermal load.

70 For new nuclear power stations over 50 megawatts, the Secretary of State for Trade and Industry is the competent authority, and the Directive has been implemented in England and Wales by the Electricity Works (Assessment of Environmental Effects) Regulations 2000 (SI No. 1927), which also applies to thermal power stations and overhead power lines (see below). Power stations of less than 50 megawatts, and other buildings to house small nuclear reactors, are

given consent through the normal planning system and are subject to the provisions described in Parts 1 and 2 of this booklet.

71 The Directive's provisions on the dismantling and decommissioning of nuclear power stations and reactors have been implemented by the Nuclear Reactors (Environmental Impact Assessment for Decommissioning) Regulations 1999 (SI No. 2892), which apply to England, Wales and Scotland. They prohibit dismantling or decommissioning from being carried out without the consent of the Health and Safety Executive. A licensee who applies for consent is required to provide the Executive with an environmental statement. The Regulations also apply to changes to existing dismantling or decommissioning projects which may have significant effects on the environment (Annex II.13 of the Directive).

Other power stations and overhead power lines

72 Annex I of the Directive includes thermal power stations with a heat output of 300 megawatts or more, and the construction of overhead power lines with a voltage of 220 kilovolts or more and a length of more than 15 kilometres. Overhead cables which are not in Annex I fall within Annex II.

73 The construction or extension of power stations exceeding 50 megawatts, and the installation of overhead power lines, require consent from the Secretary of State for Trade and Industry under sections 36 and 37 of the Electricity Act 1989, and the requirement for EIA has been implemented as part of the procedure for applications under those provisions. The Electricity Works (Environmental Impact Assessment) Regulations 2000 (SI No. 1927) require EIA for all projects which fall within Annex I of the Directive. They also require proposed power stations not covered by Annex I, and all overhead power lines of 132 kilovolts or more, to be screened for EIA. Overhead power lines below 132 kilovolts are not normally expected to have significant environmental effects and do not need to be considered for EIA unless local circumstances demand it. The Regulations also make provisions relating to nuclear power stations (see paragraph 70 above).

74 The Regulations apply to England and Wales and replace that part of the Electricity and Pipe-line Works (Assessment of Environmental

Effects) Regulations 1990 dealing with power stations and overhead cables. The 1990 Regulations, however, remain in force in respect of applications lodged before the new Regulations came into force.

75   Section 36 of the Electricity Act 1989 has been amended by the Electricity Act 1989 (Requirement of Consent for Offshore Wind and Water Driven Generating Stations) (England and Wales) Order 2001 (SI No. 3642). The Order requires the Secretary of State's consent for offshore wind and water driven generating stations with a capacity of one megawatt or more that are situated in waters within or adjacent to England and Wales up to the seaward limit of territorial seas. The Order does not apply to those projects which require planning permission under the Town and Country Planning Act 1990.

76   Power stations of 50 megawatts or less approved under planning legislation and subject to the provisions described in Parts 1 and 2 of this booklet.

Forestry projects

77   Projects in Annex II of the Directive include 'Initial afforestation and deforestation for the purposes of conversion to another type of land use.' The Environmental Impact Assessment (Forestry) (England and Wales) Regulations 1999 (SI No. 2228) require anyone who proposes to carry out a forestry project that is likely to have significant effects on the environment to apply for a consent from the Forestry Commission before starting work. Those who apply for consent will be required to prepare an environmental statement.

78   There are four categories of forestry project that fall within the Regulations: afforestation (creating new woodlands), deforestation (conversion of woodland to another use), constructing forest roads, and quarrying material to construct forest roads. The Regulations incorporate thresholds below which the project is taken to be unlikely to have significant effects on the environment. The thresholds are 5 hectares for afforestation and 1 hectare for other categories, although projects in sensitive areas have lower and in some cases no thresholds. In considering whether EIA is necessary for projects which meet the appropriate threshold, the Forestry Commission takes account of the nature and scale of the project and the sensitivity of the site.

Given the variability of forestry developments and the importance of location in determining whether significant effects on the environment are likely, it is not possible to formulate criteria which will provide a universal test of whether or not EIA is required. The question must be considered on a case-by-case basis. Anyone contemplating a forestry project is advised to contact the appropriate Forestry Commission Conservancy Office for advice on whether or not EIA will be required.

79 An applicant who disagrees with the Commission's opinion that consent is required for a project may ask (in England) the Secretary of State for the Department for Environment, Food and Rural Affairs for a direction.

80 If the Forestry Commission discovers that work is being carried out that would have required its consent or has breached the conditions of consent, it may serve an Enforcement Notice. This can require the person carrying out the work, or someone else with sufficient interest in the land, to do one or more of the following: stop the work, apply for consent, restore the land to its condition before the work was started, carry out work to secure compliance with the conditions of consent, remove or alleviate any injury to the environment which has been caused by the work.

Land drainage improvements

81 Land drainage projects fall within Annex II.1(c) of the Directive. New land drainage works, including flood defence works and defences against the sea, require planning permission and the provisions described in Parts 1 and 2 of this booklet apply to such works. Land drainage improvement works undertaken by drainage bodies are permitted development under the Town and Country Planning (General Permitted Development) Order 1995 (see paragraph 20) and are therefore exempt from planning permission. As such works might have significant effects on the environment, the principles of EIA need to be applied to them. This is done through the Environmental Impact Assessment (Land Drainage Improvement Works) Regulations 1999 (SI No. 1783), which apply to England and Wales and revoke and replace earlier Regulations which came into force in 1988 and were amended in 1995.

82   The 1999 Regulations require a drainage body to consider whether proposed improvement works are likely to have significant effects on the environment. Where the drainage body considers that there are unlikely to be such effects, it must publicise its intention to carry out the works. If the drainage body receives representations that there are likely to be significant effects but it still thinks otherwise, it must apply for a determination to the appropriate authority (Secretary of State for the Department for Environment, Food and Rural Affairs or, in Wales, the National Assembly for Wales). Where the drainage body concludes that the works are likely to have significant environmental effects, it must publicise its intention to prepare an environmental statement and notify specified consultation bodies (English Nature, the Countryside Agency and any other authority or organisation which might have an interest).

83   The environmental effects of the improvement works must be assessed in the light, in particular, of the environmental statement and any representations received. If there are no objections, the drainage body may determine that it will proceed with the works. If there are objections, the proposal must be referred to the appropriate authority for a determination giving or refusing consent to the works. The determination must be publicised.

Ports and harbours

84   Annex I.8 of the Directive includes trading ports, piers for loading and unloading connected to land and outside ports (excluding ferry piers) which can take vessels of over 1350 tonnes. Annex II.10(e) includes construction of harbours and port installations, including fishing harbours, which do not fall within Annex I.

85   Proposed development at ports and harbours down to the low water mark is subject to Town and Country Planning EIA Regulations. For other works, the requirements of the Directive have been implemented in England, Wales and Scotland by the Harbour Works (Environmental Impact Assessment) Regulations 1999 (SI No. 3445). These Regulations incorporate a revised Schedule 3 to the Harbour Works Act 1964, which gives the procedures for harbour revision orders and harbour empowerment orders, and replace the Harbour Works (Assessment of Environmental Effects) (No. 2) Regulations

1989 governing harbour works not needing to be authorised by harbour order.

86   Where the proposed works are subject to a harbour order, the Secretary of State for the Department for Transport (in the case of fishery harbours, the Secretary of State for Environment, Food and Rural Affairs) is responsible for deciding whether EIA is required and, if so, whether the works should proceed in the light of the assessment and the comments of environmental bodies.

87   In general, works under harbour empowerment orders are likely to fall within Annex I of the Directive and require EIA in every case, whereas works under harbour revision orders are likely to fall within Annex II and the need for EIA will depend on whether there are likely to be significant environmental effects. Proposed works which are subject to harbour orders are not subject to EIA if the area of works comprised in the project does not exceed 1 hectare, unless any part falls within a 'sensitive area' or the Secretary of State decides that EIA is needed in a particular case.

Marine fish farming

88   Intensive fish farming is an Annex II project. Developments of on-shore fish farming facilities may require planning permission and the provisions described in Parts 1 and 2 of this booklet apply to proposals for such facilities.

89   Off-shore facilities do not require planning permission but require a lease from the Crown Estate Commissioners (or, where appropriate, from the Shetlands Islands Council or Orkney Islands Council). The Environmental Impact Assessment (Fish Farming in Marine Waters) Regulations 1999 (SI No. 367), which apply to England, Wales and Scotland, set indicative criteria which, when triggered, require the Commissioners to screen applications to determine whether EIA is needed. They apply where the development is designed to hold a biomass of 100 tonnes or more, or where it will extend to 0.1 hectare or more of the surface area of the marine waters (including any proposed structures or excavations), or where any part of the development is to be carried out in a sensitive area. The Regulations prohibit a lease from being granted before the environmental

statement provided by the developer, and comments on it from consultees and the general public, have been considered.

90 The Crown Estate Commissioners have, in consultation with the Scottish Executive Rural Affairs Department, issued an Environmental Assessment Guidance Manual for Marine Salmon Farmers which provides guidance to developers on how to conduct EIA and prepare an environmental statement. Guidance on the practical implications of the 1999 Regulations has also been issued by the Scottish Executive Rural Affairs Department (see Appendix 8).

Marine dredging for minerals

91 Extraction of minerals by marine dredging falls within Annex II.2(c) of the Directive. The Environmental Impact Assessment and Habitats (Extraction of Minerals by Marine Dredging) Regulations, which are likely to come into force in 2004, will implement the Directive in waters around the UK (but excluding waters around Scotland). They will also implement the Habitats Directive (92/43/EEC).

92 Unless the regulator (First Secretary of State, Department for Environment Northern Ireland or the National Assembly for Wales as appropriate) has provided, in response to a request, a written determination that proposed dredging is not likely to have significant effects, application for permission must be obtained and the application must be accompanied by an environmental statement. The only exceptions to this are where the Secretary of State has determined that the dredging would form part of a project serving national defence purposes, or where the regulator has determined that it is exempted under Article 2.3 of the Directive.

Transport and Works Act 1992

93 The Transport and Works Act 1992 (TWA) enables orders to be made authorising the construction or operation of railways, tramways, other guided transport systems and inland waterways; and works interfering with rights of navigation. Orders may also authorise ancillary matters such as compulsory purchase and creating or extinguishing rights over land. Applicants for TWA orders may apply

at the same time for a direction that planning permission is deemed to be granted.

94   The procedures for making applications for orders under Part 1 of the Act are contained in the Transport and Works (Applications and Objections Procedure) (England and Wales) Rules 2000 (SI No. 2190), which also implement the Directive. These Rules apply in England and Wales and replace the earlier 1992 Rules (SI No. 2902) except for applications made and not determined before the new Rules came into force on 16 October 2000. In England, most applications for orders are determined by the Secretary of State for Transport. Applications for orders relating wholly to Wales fall to be determined by the National Assembly for Wales.

95   The procedures include provision for opposed orders to be considered by way of a public inquiry, hearing or exchanges of written representations. Schemes that are identified by the Secretary of State or the Assembly as being of national significance are also subject to Parliamentary approval prior to a public inquiry being held.

Uncultivated land and Semi-natural areas

96   The EIA (Uncultivated Land and Semi-natural Areas) Regulations 2001 (SI No. 3966) apply to the change the use of uncultivated land or semi-natural areas to intensive agricultural use. Such land is present on many farms and may include unimproved grassland, heathland, moorland, scrubland and wetlands. Examples of agricultural intensification may include ploughing, cultivation, soil spreading, drainage, reclamation, increased application of fertilisers or pesticides, and increased grazing by livestock.

97   Such projects must not commence without first obtaining a screening decision from Defra to see whether the Regulations apply and if EIA is required. Where EIA is required an application for consent has to be submitted to the Secretary of State together with an environmental statement. Projects requiring EIA must not be carried out until consent has been granted by the Secretary of State.

Separate Regulations exist for the devolved Administrations and are listed in Appendix 8.

Water management projects for agriculture

98   Water management projects for agriculture (including irrigation) fall within Annex II 1(c) of the Directive. Most such projects are likely to constitute development within the meaning of the Town and Country Planning Act 1990 and thus be subject to environmental impact assessment through the planning system. Where such a project does not constitute development and is likely to have significant effects on the environment it falls to be considered under The Water Resources (Environmental Impact Assessment) (England and Wales) Regulations 2003 (SI No.164).

99   These regulations apply in England and Wales and attach the need for formal environmental impact assessment to the water abstraction and impoundment licensing system which is administered by the Environment Agency. Projects involving the abstraction of water are included if the amounts abstracted exceed 20 cubic metres in any period of 24 hours. The regulations also make it an offence to begin or carry out a relevant project without the Environment Agency's consent in cases where no abstraction or impoundment licence is required.

Projects arising in Scotland and Northern Ireland

100 The arrangements for environmental impact assessment in Scotland and Northern Ireland are broadly similar to those applying in England and Wales, but are mainly subject to separate legislative provisions. The Environmental Impact Assessment (Scotland) Regulations 1999 (SSI 1999/1) implement the Directive in Scotland for projects which are subject to planning permission and for certain trunk road projects and land drainage works. Scottish Executive Development Department Circular 15/1999 explains the Regulations, and Planning Advice Note (PAN 58) on Environmental Impact Assessment also provides background information and guidance on best practice. The preceding section, in paragraphs 56 to 94, of this booklet indicates which of those Regulations relating to projects outside the planning system in England and Wales, also apply in Scotland.

101 The Planning (Environmental Impact Assessment) Regulations (Northern Ireland) 1999 (SR 1999/73) implement the Directive in

Northern Ireland for projects subject to planning permission. Development Control Advice Note 10 (Revised 1999) issued by the Planning Service, an agency of the Department of the Environment for Northern Ireland, explains the Regulations.

102 Appendix 8 includes a list of all implementing legislation currently in force in England and Wales, Scotland and Northern Ireland, and a list of available guidance.

APPENDIX 1

Consolidation of Directive 85/337/EEC on the assessment of the effects of certain public and private projects on the environment

As amended by Council Directive 97/11/EC (adopted by Council 3 March 1997).

(Text of Directive 97/11/EC is available in the Official Journal of the European Communities, 14 March 1997).

Text in italics is a change or an addition made by 97/11/EC.

THE COUNCIL OF THE EUROPEAN UNION

Having regard to the Treaty establishing the European Community, and in particular Article 130s (1) thereof,

Having regard to the proposal from the Commission[1],

Having regard to the opinion of the Economic and Social Committee[2],

Having regard to the opinion of the Committee of the Regions[3],

1   O.J. No. C 130, 12.5.1994, p. 8.
2   O.J. No. C 393, 31.12.1994, p. 1.
3   O.J. No. C 210, 14.8.1995, p. 78.

Acting in accordance with the procedure laid down in Article 189c of the Treaty[4],

1. *Whereas Council Directive 85/337/EEC of 27 June 1985 on the assessment of the effects of certain public and private projects on the environment[5] aims at providing the competent authorities with relevant information to enable them to take a decision on a specific project in full knowledge of the project's likely significant impact on the environment; whereas the assessment procedure is a fundamental instrument of environmental policy as defined in Article 130r of the Treaty and of the Fifth Community Programme of policy and action in relation to the environment and sustainable development;*

2. *Whereas, pursuant to Article 130r (2) of the Treaty, Community policy on the environment is based on the precautionary principle and on the principle (sic) that preventive action should be taken, that environmental damage should as a priority be rectified at source and that the polluter should pay;*

3. *Whereas the main principles of the assessment of environmental effects should be harmonised and whereas the Member States may lay down stricter rules to protect the environment;*

4. *Whereas experience acquired in environmental impact assessment, as recorded in the report on the implementation of Directive 85/337/EEC, adopted by the Commission on 2 April 1993, shows that it is necessary to introduce provisions designed to clarify, supplement and improve the rules on the assessment procedure, in order to ensure that the Directive is applied in an increasingly harmonised and efficient manner;*

5. *Whereas projects for which an assessment is required should be subject to a requirement for development consent; whereas the assessment should be carried out before such a consent is granted;*

---

4   European Parliament Opinion of 11 October 1995 (O.J. No. C 287, 30.10.1995, p. 101), Council common position of 25 June 1996 (O.J. No. C 248, 26.8.1996, p. 75) and European Parliament Decision of 13 November 1996 (O.J. No. C 362, 2.12.1996, p. 103).

5   O.J. No. L 175, 5.7.1985, p. 40. Directive as last amended by the 1994 Act of Accession.

6.   *Whereas it is appropriate to make additions to the list of projects which have significant effects on the environment and which must on that account as a rule be made subject to systematic assessment;*

7.   *Whereas projects of other types may not have significant effects on the environment in every case; whereas these projects should be assessed where Member States consider they are likely to have significant effects on the environment;*

8.   *Whereas Member States may set thresholds or criteria for the purpose of determining which such projects should be subject to assessment on the basis of the significance of their environmental effects; whereas Member States should not be required to examine projects below those thresholds or outside those criteria on a case-by-case basis;*

9.   *Whereas when setting such thresholds or criteria or examining projects on a case-by-case basis for the purpose of determining which projects should be subject to assessment on the basis of their significant environmental effects, Member States should take account of the relevant selection criteria set out in this Directive; whereas, in accordance with the subsidiarity principle, the Member States are in the best position to apply these criteria in specific instances;*

10.   *Whereas the existence of a location criterion referring to special protection areas designated by Member States pursuant to Council Directive 79/409/ EEC of 2 May 1979 on the conservation of wild birds[6] and 92/43/EEC of 21 May 1992 on the conservation of natural habitats and of wild fauna and flora[7] does not imply necessarily that projects in those areas are to be automatically subject to an assessment under this Directive;*

11.   *Whereas it is appropriate to introduce a procedure in order to enable the developer to obtain an opinion from the competent authorities on the content and the extent of the information to be elaborated and supplied for the assessment; whereas Member States, in the framework of this procedure, may require the developer to provide, inter alia, alternatives for the projects for which it intends to submit an application;*

6   O.J. No. L 103, 25.4.1979, p. 1. Directive as last amended by the 1994 Act of Accession.
7   O.J. No. L 206, 22.7.1992, p. 7.

12. *Whereas it is desirable to strengthen the provisions concerning environmental impact assessment in a transboundary context to take account of developments at international level;*

13. *Whereas the Community signed the Convention on Environmental Impact Assessment in a Transboundary Context on 25 February 1991.*

(Preamble of 85/337/EEC)
THE COUNCIL OF THE EUROPEAN COMMUNITIES

Having regard to the Treaty establishing the European Economic Community, and in particular Articles 100 and 235 thereof,

Having regard to the proposal from the Commission[1],

Having regard to the opinion of the European Parliament[2],

Having regard to the opinion of the Economic and Social Committee[3],

Whereas the 1973[4] and 1977[5] action programmes of the European Communities on the environment, as well as the 1983[6] action programme, the main outlines of which have been approved by the Council of the European Communities and the representatives of the Governments of the Member States, stress that the best environmental policy consists in preventing the creation of pollution or nuisances at source, rather than subsequently trying to counteract their effects; whereas they affirm the need to take effects on the environment into account at the earliest possible stage in all the technical planning and decision-making processes; whereas to that end, they provide for the implementation of procedures to evaluate such effects;

1   O.J. No. C 169, 9.7.1980, p. 14.
2   O.J. No. C 66, 15.3.1982, p. 89.
3   O.J. No. C 185, 27.7.1981, p. 8.
4   O.J. No. C 122, 20.12.1973, p. 1.
5   O.J. No. C 139, 13.6.1977, p. 1.
6   O.J. No. C 46, 17.2.1983, p. 1.

Whereas the disparities between the laws in force in the various Member States with regard to the assessment of the environmental effects of public and private projects may create unfavourable competitive conditions and thereby directly affect the functioning of the common market; whereas, therefore, it is necessary to approximate national laws in this field pursuant to Article 100 of the Treaty;

Whereas, in addition, it is necessary to achieve one of the Community's objectives in the sphere of the protection of the environment and the quality of life;

Whereas, since the Treaty has not provided the powers required for this end, recourse should be had to Article 235 of the Treaty;

Whereas general principles for the assessment of environmental effects should be introduced with a view to supplementing and coordinating development consent procedures governing public and private projects likely to have a major effect on the environment;

Whereas development consent for public and private projects which are likely to have significant effects on the environment should be granted only after prior assessment of the likely significant environmental effects of these projects has been carried out; whereas this assessment must be conducted on the basis of the appropriate information supplied by the developer, which may be supplemented by the authorities and by the people who may be concerned by the project in question;

Whereas the principles of the assessment of environmental effects should be harmonised, in particular with reference to the projects which should be subject to assessment, the main obligations of the developers and the content of the assessment;

Whereas projects belonging to certain types have significant effects on the environment and these projects must as a rule be subject to systematic assessment;

Whereas projects of other types may not have significant effects on the environment in every case and whereas these projects should be assessed where the Member States consider that their characteristics so require;

Whereas, for projects which are subject to assessment, a certain minimal amount of information must be supplied, concerning the project and its effects;

Whereas the effects of a project on the environment must be assessed in order to take account of concerns to protect human health, to contribute by means of a better environment to the quality of life, to ensure maintenance of the diversity of species and to maintain the reproductive capacity of the ecosystem as a basic resource for life;

Whereas, however, this Directive should not be applied to projects the details of which are adopted by a specific act of national legislation, since the objectives of this Directive, including that of supplying information, are achieved through the legislative process;

Whereas, furthermore, it may be appropriate in exceptional cases to exempt a specific project from the assessment procedures laid down by this Directive, subject to appropriate information being supplied to the Commission.

HAS ADOPTED THIS DIRECTIVE:

Article 1

1. This Directive shall apply to the assessment of the environmental effects of those public and private projects which are likely to have significant effects on the environment.

2. For the purposes of this Directive:

'project' means:

the execution of construction works or of other installations or schemes,
other interventions in the natural surroundings and landscape including those involving the extraction of mineral resources;

'developer' means:

the applicant for authorisation for a private project or the public authority which initiates a project;

'development consent' means:

the decision of the competent authority or authorities which entitled the developer to proceed with the project.

3. The competent authority or authorities shall be that or those which the Member States designate as responsible for performing the duties arising from this Directive.

4. Projects serving national defence purposes are not covered by this Directive.

5. This Directive shall not apply to projects the details of which are adopted by a specific act of national legislation, since the objectives of this Directive, including that of supplying information, are achieved through the legislative process.

Article 2

1. Member States shall adopt all measures necessary to ensure that, before consent is given, projects likely to have significant effects on the environment by virtue, inter alia, of their nature, size or location are made subject to a *requirement for development consent and* an assessment with regard to their effects. These projects are defined in Article 4.

2. The environmental impact assessment may be integrated into the existing procedures for consent to projects in the Member States, or, failing this, into other procedures or into procedures to be established to comply with the aims of this Directive.

2a. *Member States may provide a single procedure in order to fulfil the requirements of this Directive and the requirements of Council Directive 96/61/EC of 24 September 1996 on integrated pollution prevention and control[7].*

7   O.J. No. L 257, 10.10.1996, p. 26.

3. *Without prejudice to Article 7*, Member States may, in exceptional cases, exempt a specific project in whole or in part from the provisions laid down in this Directive.

    In this event, the Member States shall:

    (a) consider whether another form of assessment would be appropriate and whether the information thus collected should be made available to the public;
    (b) make available to the public concerned, the information relating to the exemption and the reasons for granting it;
    (c) inform the Commission, prior to granting consent, of the reasons justifying the exemption granted, and provide it with the information made available, where *applicable* to their own nationals.

    The Commission shall immediately forward the documents received to the other Member States.

    The Commission shall report annually to the Council on the application of this paragraph.

Article 3

The environmental impact assessment shall identify, describe and assess in an appropriate manner, in the light of each individual case and in accordance with Articles 4 to 11, the direct and indirect effects of a project on the following factors:

> human beings, fauna and flora;
> soil, water, air, climate and the landscape;
> material assets and the cultural heritage;
> the interaction between the factors mentioned in the first, second *and third* indents.

Article 4

1. Subject to Article 2(3), projects listed in Annex I shall be made subject to an assessment in accordance with Articles 5 to 10.

2. *Subject to Article 2(3), for projects listed in Annex II, the Member States shall determine through:*

   (a) a case-by-case examination; or

   (b) thresholds or criteria set by the Member State

   whether the project shall be made subject to an assessment in accordance with Articles 5 to 10.

   Member States may decide to apply both procedures referred to in (a) and (b).

3. *When a case-by-case examination is carried out or thresholds or criteria are set for the purpose of paragraph 2, the relevant selection criteria set out in Annex III shall be taken into account.*

4. *Member States shall ensure that the determination made by the competent authorities under paragraph 2 is made available to the public.*

Article 5

1. In the case of projects which, pursuant to Article 4, must be subjected to an environmental impact assessment in accordance with Articles 5 to 10, Member States shall adopt the necessary measures to ensure that the developer supplies in an appropriate form the information specified in Annex IV inasmuch as:

   (a) the Member States consider that the information is relevant to a given stage of the consent procedure and to the specific characteristics of a particular project or type of project and of the environmental features likely to be affected;

   (b) the Member States consider that a developer may reasonably be required to compile this information having regard inter alia to current knowledge and methods of assessment.

2. *Member States shall take the necessary measures to ensure that, if the developer so requests before submitting an application for development consent, the competent authority shall give an opinion on the information to be supplied by the developer in accordance with paragraph 1. The competent authority shall consult the developer and authorities referred to in Article 6(1) before it gives its opinion.*

The fact that the authority has given an opinion under this paragraph shall not preclude it from subsequently requiring the developer to submit further information.

Member States may require the competent authorities to give such an opinion, irrespective of whether the developer so requests.

3. The information to be provided by the developer in accordance with paragraph 1 shall include at least:

    a description of the project comprising information on the site, design and size of the project;
    a description of the measures envisaged in order to avoid, reduce and, if possible, remedy significant adverse effects;
    the data required to identify and assess the main effects which the project is likely to have on the environment;
    an outline of the main alternatives studied by the developer and an indication of the main reasons for his choice, taking into account the environmental effects;
    a non-technical summary of the information mentioned in the *previous* indents.

4. Member States shall, if necessary, ensure that any authorities holding relevant information, *with particular reference to Article 3*, shall make this information available to the developer.

Article 6

1. Member States shall take the measures necessary to ensure that the authorities likely to be concerned by the project by reason of their specific environmental responsibilities are given an opportunity to express their opinion on *the information supplied by the developer and* on the request for development consent. *To this end*, Member States shall designate the authorities to be consulted, either in general terms or on a *case-by-case basis*. The information gathered pursuant to Article 5 shall be forwarded to those authorities. Detailed arrangements for consultation shall be laid down by the Member States.

2. Member States shall ensure that any request for development consent and any information gathered pursuant to Article 5 are made

available to the public *within a reasonable time, in order to give* the public concerned the opportunity to express an opinion *before the development consent is granted.*

3. The detailed arrangements for such information and consultation shall be determined by the Member States, which may in particular, depending on the particular characteristics of the projects or sites concerned:

> determine the public concerned;
> specify the places where the information can be consulted;
> specify the way in which the public may be informed, for example, by bill-posting within a certain radius, publication in local newspapers, organisation of exhibitions with plans, drawings, tables, graphs, models;
> determine the manner in which the public is to be consulted, for example, by written submissions, by public enquiry;
> fix appropriate time limits for the various stages of the procedure in order to ensure that a decision is taken within a reasonable period.

Article 7

1. Where a Member State is aware that a project is likely to have significant effects on the environment in another Member State or where a Member State likely to be significantly affected so requests, the Member State in whose territory the project is intended to be carried out *shall send to the affected Member State as soon as possible and no later than when informing its own public, inter alia:*

    (a) a description of the project, together with any available information on its possible transboundary impact;
    (b) information on the nature of the decision which may be taken,

    and shall give the other Member State a reasonable time in which to indicate whether it wishes to participate in the Environmental Impact Assessment procedure, and may include the information referred to in paragraph 2.

2. *If a Member State which receives information pursuant to paragraph 1 indicates that it intends to participate in the Environmental Impact*

Assessment procedure, the Member State in whose territory the project is intended to be carried out shall, if it has not already done so, send to the affected Member State the information gathered pursuant to Article 5 and relevant information regarding the said procedure, including the request for development consent.

3. *The Member States concerned, each insofar as it is concerned, shall also:*

    (a) *arrange for the information referred to in paragraphs 1 and 2 to be made available, within a reasonable time, to the authorities referred to in Article 6(1) and the public concerned in the territory of the Member State likely to be significantly affected; and*

    (b) *ensure that those authorities and the public concerned are given an opportunity, before development consent for the project is granted, to forward their opinion within a reasonable time on the information supplied to the competent authority in the Member State in whose territory the project is intended to be carried out.*

4. *The Member States concerned shall enter into consultations concerning, inter alia, the potential transboundary effects of the project and the measures envisaged to reduce or eliminate such effects and shall agree on a reasonable timeframe for the duration of the consultation period.*

5. *The detailed arrangements for implementing the provisions of this Article may be determined by the Member States concerned.*

Article 8

The results of consultations and the information gathered pursuant to Articles 5, 6 and 7 must be taken into consideration in the development consent procedure.

Article 9

1. When a decision *to grant or refuse development consent* has been taken, the competent authority or authorities shall inform the public thereof *in accordance with the appropriate procedures and shall make available to the public the following information:*

    the content of the decision and any conditions attached thereto;

the *main* reasons and considerations on which the decision is based;

a description, where necessary, of the main measures to avoid, reduce and, if possible, offset the major adverse effects.

2.   The competent authority or authorities shall inform any Member State which has been consulted pursuant to Article 7, forwarding to it the information referred to in paragraph 1.

Article 10

The provisions of this Directive shall not affect the obligation on the competent authorities to respect the limitations imposed by national regulations and administrative provisions and accepted legal practices with regard to *commercial and industrial confidentiality, including intellectual property*, and the safeguarding of the public interest.

Where Article 7 applies, the transmission of information to another Member State and the receipt of information by another Member State shall be subject to the limitations in force in the Member State in which the project is proposed.

Article 11

1.   The Member States and the Commission shall exchange information on the experience gained in applying this Directive.

2.   In particular, Member States shall inform the Commission of any criteria and/or thresholds adopted for the selection of the projects in question, in accordance with Article 4(2).

3.   Five years after notification of this Directive, the Commission shall send the European Parliament and the Council a report on its application and effectiveness. The report shall be based on the aforementioned exchange of information.

4.   On the basis of this exchange of information, the Commission shall submit to the Council additional proposals, should this be necessary,

with a view to this Directive's being applied in a sufficiently coordinated manner.

Article 12

1. Member States shall take the measures necessary to comply with this Directive [*i.e.* 85/337/EEC] within three years of its notification.

2. Member States shall communicate to the Commission the texts of the provisions of national law which they adopt in the field covered by this Directive.

Article 2 of 97/11/EC

Five years after the entry into force of this Directive, the Commission shall send the European Parliament and the Council a report on the application and effectiveness of Directive 85/337/EEC as amended by this Directive. The report shall be based on the exchange of information provided for by Article 11(1) and (2).

On the basis of this report, the Commission shall, where appropriate, submit to the Council additional proposals with a view to ensuring further coordination in the application of this Directive.

Article 3 of 97/11/EC

1. *Member States shall bring into force the laws, regulations and administrative provisions necessary to comply with this Directive by 14 March 1999 at the latest. They shall inform the Commission thereof.*

   When Member States adopt these provisions, these shall contain a reference to this Directive or shall be accompanied by such reference at the time of their official publication. The procedure for such reference shall be adopted by Member States.

2. *If a request for development consent is submitted to a competent authority before the end of the time limit laid down in paragraph 1, the provisions of Directive 85/337/EEC prior to these amendments shall continue to apply.*

Article 4 of 97/11/EC

This Directive shall enter into force on the twentieth day following that of its publication in the Official Journal of the European Communities.

Article 5 of 97/11/EC

This Directive is addressed to the Member States.

Done at Brussels,   *For the Council*
                                                           The President

ANNEX I
Projects subject to Article 4(1)

1. Crude-oil refineries (excluding undertakings manufacturing only lubricants from crude oil) and installations for the gasification and liquefaction of 500 tonnes or more of coal or bituminous shale per day.

2. Thermal power stations and other combustion installations with a heat output of 300 megawatts or more, and
   nuclear power stations and other nuclear reactors *including the dismantling or decommissioning of such power stations or reactors*[1] (except research installations for the production and conversion of fissionable and fertile material, whose maximum power does not exceed 1 kilowatt continuous thermal load).

3. (a) *Installations for the reprocessing of irradiated nuclear fuel.*

   (b) Installations designed:

   for the production or enrichment of nuclear fuel,
   for the processing of irradiated nuclear fuel or high-level radioactive waste,
   for the final disposal of irradiated nuclear fuel,
   solely for the final disposal of radioactive waste,
   solely for the storage (planned for more than 10 years) of irradiated nuclear fuels or radioactive waste in a different site than the production site.

4. Integrated works for the initial smelting of cast-iron and steel;

---

1 Nuclear power stations and other nuclear reactors cease to be such an installation when all nuclear fuel and other radioactively contaminated elements have been removed permanently from the installation site.

installations for the production of non-ferrous crude metals from ore, concentrates or secondary raw materials by metallurgical, chemical or electrolytic processes.

5.   Installations for the extraction of asbestos and for the processing and transformation of asbestos and products containing asbestos: for asbestos-cement products, with an annual production of more than 20 000 tonnes of finished products, for friction material, with an annual production of more than 50 tonnes of finished products, and for other uses of asbestos, utilisation of more than 200 tonnes per year.

6.   Integrated chemical installations, *i.e. those installations for the manufacture on an industrial scale of substances using chemical conversion processes, in which several units are juxtaposed and are functionally linked to one another and which are:*

(i)   *for the production of basic organic chemicals;*
(ii)   *for the production of basic inorganic chemicals;*
(iii)   *for the production of phosphorous-, nitrogen- or potassium-based fertilizers (simple or compound fertilizers);*
(iv)   *for the production of basic plant health products and of biocides;*
(v)   *for the production of basic pharmaceutical products using a chemical or biological process;*
(vi)   *for the production of explosives.*

7.   (a) Construction of lines for long-distance railway traffic and of airports[2] with a basic runway length of 2100 metres or more;
      (b) construction of motorways and express roads[3];
      (c) *construction of a new road of four or more lanes, or realignment and/ or widening of an existing road of two lanes or less so as to provide four or more lanes, where such new road, or realigned and/or widened section of road would be 10 kilometres or more in a continuous length.*

---

2   For the purposes of this Directive, 'airport' means airports which comply with the definition in the 1944 Chicago Convention setting up the International Civil Aviation Organisation (Annex 14).

3   For the purposes of the Directive, 'express road' means a road which complies with the definition in the European Agreement on Main International Traffic Arteries of 15 November 1975.

8. (a) Inland waterways and ports for inland-waterway traffic which permit the passage of vessels of over 1350 tonnes;

    (b) trading ports, *piers for loading and unloading connected to land and outside ports (excluding ferry piers)* which can take vessels over 1350 tonnes.

9. Waste disposal installations for the incineration, chemical treatment *as defined in Annex IIA to Directive 75/442/EEC*[4] *under heading D9, or landfill of hazardous waste (i.e. waste to which Directive 91/689/EEC*[5] *applies).*

10. *Waste disposal installations for the incineration or chemical treatment as defined in Annex IIA to Directive 75/442/EEC under heading D9 of non-hazardous waste with a capacity exceeding 100 tonnes per day.*

11. *Groundwater abstraction or artificial groundwater recharge schemes where the annual volume of water abstracted or recharged is equivalent to or exceeds 10 million cubic metres.*

12. (a) *Works for the transfer of water resources between river basins where this transfer aims at preventing possible shortages of water and where the amount of water transferred exceeds 100 million cubic metres per year;*

    (b) *in all other cases, works for the transfer of water resources between river basins where the multi-annual average flow of the basin of abstraction exceeds 2000 million cubic metres per year and where the amount of water transferred exceeds 5% of this flow.*

    In both cases transfers of piped drinking water are excluded.

13. *Waste water treatment plants with a capacity exceeding 150 000 population equivalent as defined in Article 2 point (6) of Directive 91/271/EEC.*[6]

4   O.J. No. L 194, 25.7.1975, p. 39. Directive as last amended by Commission Decision 94/3/EC (O.J. No. L 5, 7.1.1994, p. 15).

5   O.J. No. L 337, 31.12.1991, p. 20. Directive as last amended by Directive 94/31/EC (O.J. No. L 168, 2.7.1994, p. 28).

6   O.J. No. L 135, 30.5.1991, p. 40. Directive as last amended by the 1994 Act of Accession.

14.   *Extraction of petroleum and natural gas for commercial purposes where the amount extracted exceeds 500 tonnes per day in the case of petroleum and 500 000 cubic metres per day in the case of gas.*

15.   *Dams and other installations designed for the holding back or permanent storage of water, where a new or additional amount of water held back or stored exceeds 10 million cubic metres.*

16.   *Pipe-lines for the transport of gas, oil or chemicals with a diameter of more than 800 millimetres and a length of more than 40 kilometres.*

17.   *Installations for the intensive rearing of poultry or pigs with more than:*

   (a) 85 000 places for broilers, 60 000 places for hens;
   (b) 3000 places for production pigs (over 30 kilograms); or
   (c) 900 places for sows.

18.   *Industrial plants for*

   (a) the production of pulp from timber or similar fibrous materials;
   (b) the production of paper and board with a production capacity exceeding 200 tonnes per day.

19.   *Quarries and open-cast mining where the surface of the site exceeds 25 hectares, or peat extraction, where the surface of the site exceeds 150 hectares.*

20.   *Construction of overhead electrical power lines with a voltage of 220 kilovolts or more and a length of more than 15 kilometres.*

21.   *Installations for storage of petroleum, petrochemical or chemical products with a capacity of 200 000 tonnes or more.*

ANNEX II
Projects subject to Article 4(2)

1. Agriculture, silviculture and aquaculture

   (a) projects for the restructuring of rural land holdings;
   (b) projects for the use of uncultivated land or semi-natural areas for intensive agricultural purposes;
   (c) water management projects for agriculture, *including irrigation and land drainage projects;*
   (d) initial afforestation *and deforestation* for the purposes of conversion to another type of land use;
   (e) *intensive livestock installations (projects not included in Annex I);*
   (f) *intensive fish farming;*
   (g) reclamation of land from the sea.

2. Extractive industry

   (a) *quarries, open-cast mining* and peat extraction (*projects not included in Annex I);*
   (b) *underground mining;*
   (c) *extraction of minerals by marine or fluvial dredging;*
   (d) deep drillings, in particular:

   geothermal drilling,
   drilling for the storage of nuclear waste material,
   drilling for water supplies,

   with the exceptions of drillings for investigating the stability of the soil;
   (e) surface industrial installations for the extraction of coal, petroleum, natural gas and ores, as well as bituminous shale.

3.   Energy industry

   (a) industrial installation for the production of electricity, steam and
        hot water *(projects not included in Annex I)*;
   (b) industrial installations for carrying gas, steam and hot water;
        transmission of electrical energy by overhead cables *(projects not
        included in Annex I)*;
   (c) surface storage of natural gas;
   (d) underground storage of combustible gases;
   (e) surface storage of fossil fuels;
   (f) industrial briquetting of coal and lignite;
   (g) installations for the processing and storage of radioactive waste
        (unless included in Annex I);
   (h) installations for hydroelectric energy production;
   (i) *installations for the harnessing of wind power for energy production
        (wind farms).*

4.   Production and processing of metals

   (a) installations for the production of pig iron or steel (primary or
        secondary fusion) *including continuous casting;*
   (b) installations for the processing of ferrous metals:
        (i)   hot-rolling mills;
        (ii) *smitheries with hammers;*
        (iii) *application of protective fused metal coats.*

   (c) ferrous metal foundries;
   (d) installations for the smelting, *including the alloyage,* of non-ferrous
        metals, excluding precious metals, *including recovered products*
        (refining, foundry casting, etc.);
   (e) installations for surface treatment of metals *and plastic materials
        using an electrolytic or chemical process;*
   (f) manufacture and assembly of motor vehicles and manufacture of
        motor-vehicle engines;
   (g) shipyards;
   (h) installations for the construction and repair of aircraft;
   (i) manufacture of railway equipment;
   (j) swaging by explosives;
   (k) installations for the roasting and sintering of metallic ores.

5. Mineral industry

    (a) coke ovens (dry coal distillation);

    (b) installations for the manufacture of cement;

    (c) installations for the production of asbestos *and the manufacture of asbestos-based products (projects not included in Annex I)*;

    (d) installations for the manufacture of glass including glass fibre;

    (e) *installations for smelting mineral substances including the production of mineral fibres;*

    (f) *manufacture of ceramic products by burning, in particular roofing tiles, bricks, refractory bricks, tiles, stoneware or porcelain.*

6. Chemical industry (projects not included in Annex I)

    (a) treatment of intermediate products and productions of chemicals;

    (b) production of pesticides and pharmaceutical products, paint and varnishes, elastomers and peroxides;

    (c) storage facilities for petroleum, petrochemical and chemical products.

7. Food industry

    (a) manufacture of vegetable and animal oils and fats;

    (b) packing and canning of animal and vegetable products;

    (c) manufacture of dairy products;

    (d) brewing and malting;

    (e) confectionery and syrup manufacture;

    (f) installations for the slaughter of animals;

    (g) industrial starch manufacturing installations;

    (h) fish-meal and fish-oil factories;

    (i) sugar factories.

8. Textile, leather, wood and paper industries

    (a) *industrial plants for the production of* paper and board *(projects not included in Annex I)*;

    (b) plants for the pretreatment (operations such as washing, bleaching, mercerisation) or dyeing of fibres or textiles;

    (c) *plants for the* tanning *of hides and skins;*

    (d) cellulose-processing and production installations.

9.   Rubber industry

Manufacture and treatment of elastomer-based products.

10.   Infrastructure projects

   (a)   industrial estate development projects;
   (b)   urban development projects, *including the construction of shopping centres and car parks*;
   (c)   *construction of railways and intermodal transshipment facilities, and of intermodal terminals (projects not included in Annex I)*;
   (d)   construction of airfields (projects not included in Annex I);
   (e)   construction of roads, harbours *and port installations*, including fishing harbours (projects not included in Annex I);
   (f)   *inland-waterway construction not included in Annex I*, canalisation and flood-relief works;
   (g)   dams and other installations designed to hold water or store it on a long-term basis (*projects not included in Annex I*);
   (h)   tramways, elevated and underground railways, suspended lines or similar lines of a particular type, used exclusively or mainly for passenger transport;
   (i)   oil and gas pipeline installations (*projects not included in Annex I*);
   (j)   installation of long-distance aqueducts;
   (k)   *coastal work to combat erosion and maritime works capable of altering the coast through the construction, for example, of dykes, moles, jetties and other sea defence works, excluding the maintenance and reconstruction of such works*;
   (l)   *groundwater abstraction and artificial groundwater recharge schemes not included in Annex I*;
   (m)   *works for the transfer of water resources between river basins not included in Annex I.*

11.   Other projects

   (a)   permanent racing and test tracks for *motorised vehicles*;
   (b)   installations for the disposal of *waste* (projects not included in Annex I);
   (c)   waste water treatment plants (*projects not included in Annex I*);
   (d)   sludge-deposition sites;
   (e)   storage of scrap iron, *including scrap vehicles*;
   (f)   test benches for engines, turbines or reactors;

(g) *installations for the manufacture* of artificial mineral fibres;

(h) *installations for the recovery or destruction of explosive substances*;

(i) knackers' yards.

12. Tourism and leisure

(a) *ski-runs, ski-lifts and cable-cars and associated developments*;

(b) *marinas*;

(c) holiday villages and hotel complexes *outside urban areas and associated developments*;

(d) *permanent camp sites and caravan sites*;

(e) *theme parks*.

13. *Any change or extension of projects listed in Annex I or Annex II, already authorised, executed or in the process of being executed, which may have significant adverse effects on the environment;*
projects in Annex I, undertaken exclusively or mainly for the development and testing of new methods or products and not used for more than *two* years.

ANNEX III
Selection criteria referred to in Article 4(3)

1. Characteristics of projects

The characteristics of projects must be considered having regard, in particular, to:

the size of the project;
the cumulation with other projects;
the use of natural resources;
the production of waste;
pollution and nuisances;
the risk of accidents, having regard in particular to substances or technologies used.

2. Location of projects

The environmental sensitivity of geographical areas likely to be affected by projects must be considered, having regard, in particular, to:

the existing land use;
the relative abundance, quality and regenerative capacity of natural resources in the area;
the absorption capacity of the natural environment, paying particular attention to the following areas:

(a) wetlands;
(b) coastal zones;
(c) mountain and forest areas;
(d) nature reserves and parks;
(e) areas classified or protected under Member States' legislation; special protection areas designated by Member States pursuant to Directive 79/409/EEC and 92/43/EEC;

(f) areas in which the environmental quality standards laid down in Community legislation have already been exceeded;

(g) densely populated areas;

(h) landscapes of historical, cultural or archaeological significance.

3. Characteristics of the potential impact

The potential significant effects of projects must be considered in relation to criteria set out under 1 and 2 above, and having regard in particular to:

the extent of the impact (geographical area and size of the affected population);

the transfrontier nature of the impact;

the magnitude and complexity of the impact;

the probability of the impact;

the duration, frequency and reversibility of the impact.

ANNEX IV
Information referred to in
Article 5(1)

1.   Description of the project, including in particular:

a description of the physical characteristics of the whole project and the land-use requirements during the construction and operational phases;

a description of the main characteristics of the production processes, for instance, nature and quantity of the materials used;

an estimate, by type and quantity, of expected residues and emissions (water, air and soil pollution, noise, vibration, light, heat, radiation, etc.) resulting from the operation of the proposed project.

2.   An *outline* of the main alternatives studied by the developer and an indication of the main reasons for his choice, taking into account the environmental effects.

3.   A description of the aspects of the environment likely to be significantly affected by the proposed project, including, in particular, population, fauna, flora, soil, water, air, climatic factors, material assets, including the architectural and archaeological heritage, landscape and the inter-relationship between the above factors.

4.   A description[1] of the likely significant effects of the proposed project on the environment resulting from:

---

1   This description should cover the direct effects and any indirect, secondary, cumulative, short, medium and long-term, permanent and temporary, positive and negative effects of the project.

the existence of the project;
the use of natural resources;
the emission of pollutants, the creation of nuisances and the
elimination of waste;

and the description by the developer of the forecasting methods used
to assess the effects on the environment.

5. A description of the measures envisaged to prevent, reduce and
    where possible offset any significant adverse effects on the
    environment.

6. A non-technical summary of the information provided under the
    above headings.

7. An indication of any difficulties (technical deficiencies or lack of
    know-how) encountered by the developer in compiling the required
    information.

APPENDIX 2
Projects to which the Town and Country Planning (Environmental Impact Assessment) (England and Wales) Regulations 1999 apply: Schedule 1 projects

The following types of development require environmental impact assessment in every case:

1. Crude-oil refineries (excluding undertakings manufacturing only lubricants from crude oil) and installations for the gasification and liquefaction of 500 tonnes or more of coal or bituminous shale per day.

2. (a) Thermal power stations and other combustion installations with a heat output of 300 megawatts or more; and
   (b) nuclear power stations and other nuclear reactors[1] (except research installations for the production and conversion of fissionable and fertile materials, whose maximum power does not exceed 1 kilowatt continuous thermal load).

---

1   'nuclear power station' and 'other nuclear reactor' do not include an installation from the site of which all nuclear fuel and other radioactive contaminated materials have been permanently removed; and development for the purpose of dismantling or decommissioning a nuclear power station or other nuclear reactor shall not be treated as development of the description mentioned in paragraph 2(b).

3. (a) Installations for the reprocessing of irradiated nuclear fuel.
   (b) Installations designed:

   (i) for the production or enrichment of nuclear fuel,
   (ii) for the processing of irradiated nuclear fuel or high-level radioactive waste,
   (iii) for the final disposal of irradiated nuclear fuel,
   (iv) solely for the final disposal of radioactive waste,
   (v) solely for the storage (planned for more than ten years) of irradiated nuclear fuels or radioactive waste in a different site than the production site.

4. (a) Integrated works for the initial smelting of cast-iron and steel;
   (b) installations for the production of non-ferrous crude metals from ore, concentrates or secondary raw materials by metallurgical, chemical or electrolytic processes.

5. Installations for the extraction of asbestos and for the processing and transformation of asbestos and products containing asbestos:

   (a) for asbestos-cement products, with an annual production of more than 20 000 tonnes of finished products;
   (b) for friction material, with an annual production of more than 50 tonnes of finished products; and
   (c) for other uses of asbestos, utilisation of more than 200 tonnes per year.

6. Integrated chemical installations, that is to say, installations for the manufacture on an industrial scale of substances using chemical conversion processes, in which several units are juxtaposed and are functionally linked to one another and which are:

   (a) for the production of basic organic chemicals;
   (b) for the production of basic inorganic chemicals;
   (c) for the production of phosphorous-, nitrogen- or potassium-based fertilizers (simple or compound fertilizers);
   (d) for the production of basic plant health products and biocides;
   (e) for the production of basic pharmaceutical products using a chemical or biological process;
   (f) for the production of explosives.

7. (a) Construction of lines for long-distance railway traffic and of airports[2] with a basic runway length of 2100 metres or more;
    (b) construction of motorways and express roads[3];
    (c) construction of a new road of four or more lanes, or realignment and/or widening of an existing road of two lanes or less so as to provide four or more lanes, where such new road, or realigned and/or widened section of road would be 10 kilometres or more in a continuous length.

8. (a) Inland waterways and ports for inland-waterway traffic which permit the passage of vessels of over 1350 tonnes;
    (b) trading ports, piers for loading and unloading connected to land and outside ports (excluding ferry piers) which can take vessels of over 1350 tonnes.

9. Waste disposal installations for the incineration, chemical treatment (as defined in Annex IIA to Council Directive 75/442/EEC[4] under heading D9), or landfill of hazardous waste (that is to say, waste to which Council Directive 91/689/EEC[5] applies).

10. Waste disposal installations for the incineration or chemical treatment (as defined in Annex IIA to Council Directive 75/442/EEC under heading D9) of non-hazardous waste with a capacity exceeding 100 tonnes per day.

11. Groundwater abstraction or artificial groundwater recharge schemes where the annual volume of water abstracted or recharged is equivalent to or exceeds 10 million cubic metres.

2   'airport' means an airport which complies with the definition in the 1944 Chicago Convention setting up the International Civil Aviation Organisation (Annex 14).
3   'express road' means a road which complies with the definition in the European Agreement on Main International Traffic Arteries of 15 November 1975.
4   O.J. No. L194, 25.7.1975, p. 39. Council Directive 75/442/EEC was amended by Council Directive 91/156/EEC (O.J. No. L 78, 26.3.1991, p. 32) and by Commission Decision 94/3/EC (O.J. No. L 5, 7.1.1994, p. 15).
5   O.J. No. L 337, 31.12.1991, p. 20. Council Directive 91/689/EEC was amended by Council Directive 94/31/EC (O.J. No. L 168, 2.7.1994, p. 28).

12. (a) Works for the transfer of water resources, other than piped drinking water, between river basins where the transfer aims at preventing possible shortages of water and where the amount of water transferred exceeds 100 million cubic metres per year;

    (b) in all other cases, works for the transfer of water resources, other than piped drinking water, between river basins where the multi-annual average flow of the basin of abstraction exceeds 2000 million cubic metres per year and where the amount of water transferred exceeds 5% of this flow.

13. Waste water treatment plants with a capacity exceeding 150 000 population equivalent as defined in Article 2 point (6) of Council Directive 91/271/EEC[6].

14. Extraction of petroleum and natural gas for commercial purposes where the amount extracted exceeds 500 tonnes per day in the case of petroleum and 500 000 cubic metres per day in the case of gas.

15. Dams and other installations designed for the holding back or permanent storage of water, where a new or additional amount of water held back or stored exceeds 10 million cubic metres.

16. Pipelines for the transport of gas, oil or chemicals with a diameter of more than 800 millimetres and a length of more than 40 kilometres.

17. Installations for the intensive rearing of poultry or pigs with more than:

    (a) 85 000 places for broilers or 60 000 places for hens;
    (b) 3000 places for production pigs (over 30 kilograms); or
    (c) 900 places for sows.

18. Industrial plants for:

    (a) the production of pulp from timber or similar fibrous materials;
    (b) the production of paper and board with a production capacity exceeding 200 tonnes per day.

---

6   O.J. No. L 135, 30.5.1991, p. 40.

19. Quarries and open-cast mining where the surface of the site exceeds 25 hectares, or peat extraction where the surface of the site exceeds 150 hectares.

20. Installations for storage of petroleum, petrochemical or chemical products with a capacity of 200 000 tonnes or more.

APPENDIX 3

Projects to which the Town and Country Planning (Environmental Impact Assessment) (England and Wales) Regulations 1999 apply: Schedule 2 projects

1. In the table below:

    'area of the works' includes any area occupied by apparatus, equipment, machinery, materials, plant, spoil heaps or other facilities or stores required for construction or installation;
    'controlled waters' has the same meaning as in the Water Resources Act 1991[1];
    'floorspace' means the floorspace in a building or buildings.

2. The table below sets out the descriptions of development and applicable thresholds and criteria for the purpose of classifying development as Schedule 2 development.

---

1   1991 c. 57. *See* section 104.

Table

The carrying out of development to provide any of the following:

1. Agriculture and aquaculture

Column 1 Description of development	Column 2 Applicable thresholds and criteria	Column 3 Indicative thresholds and criteria
(a) Projects for the use of uncultivated land or semi-natural areas for intensive agricultural purposes;	The area of the development exceeds 0.5 hectare.	Development (such as greenhouses, farm buildings, etc.) on previously uncultivated land is unlikely to require EIA unless it covers more than 5 hectares. In considering whether particular development is likely to have significant effects, consideration should be given to impacts on the surrounding ecology, hydrology and landscape.
(b) Water management projects for agriculture, including irrigation and land drainage projects;	The area of the works exceeds 1 hectare.	EIA is more likely to be required if the development would result in permanent changes to the character of more than 5 hectares of land. In assessing the significance of any likely effects, particular regard should be had to whether the development would have damaging wider impacts on hydrology and surrounding ecosystems. It follows that EIA will not normally be required for routine water management projects undertaken by farmers.
(c) Intensive livestock installations (unless included in Schedule 1);	The area of new floorspace exceeds 500 square metres.	The significance or otherwise of the impacts of intensive livestock installations will often depend on the level of odours, increased traffic and the arrangements for waste handling. EIA is more likely to be required for intensive livestock installations if they are designed to house more than 750 sows, 2000 fattening pigs, 60 000 broilers or 50 000 layers, turkeys or other poultry.
(d) Intensive fish farming;	The installation resulting from the development is designed to produce more than 10 tonnes of dead weight fish per year.	Apart from the physical scale of any development, the likelihood of significant effects will generally depend on the extent of any likely wider impacts on the hydrology and ecology of the surrounding area. Developments designed to produce more than 100 tonnes (dead weight) of fish per year will be more likely to require EIA.

(e) Reclamation of land from the sea.	All development.	In assessing the significance of any development, regard should be had to the likely wider impacts on natural coastal processes beyond the site itself, as well as to the scale of reclamation works themselves. EIA is more likely to be required where work is proposed on a site which exceeds 1 hectare.

2. Extractive industry

(a) Quarries, open-cast mining and peat extraction (unless included in Schedule 1); (b) Underground mining;	All developments except the construction of buildings or other ancillary structures where the new floorspace does not exceed 1000 square metres.	The likelihood of significant effects will tend to depend on the scale and duration of the works, and the likely consequent impact of noise, dust, discharges to water and visual intrusion. All new open cast mines and underground mines will generally require EIA. For clay, sand and gravel workings, quarries and peat extraction sites, EIA is more likely to be required if they would cover more than 15 hectares or involve the extraction of more than 30 000 tonnes of mineral per year.
(c) Extraction of minerals by fluvial dredging;	All development.	Particular consideration should be given to noise, and any wider impacts on the surrounding hydrology and ecology. EIA is more likely to be required where it is expected that more than 100 000 tonnes of mineral will be extracted per year.
(d) Deep drillings, in particular: (i) geothermal drilling; (ii) drilling for the storage of nuclear waste material; (iii) drilling for water supplies; with the exception of drillings for investigating the stability of the soil.	(i) In relation to any type of drilling, the area of the works exceeds 1 hectare; or (ii) in relation to geothermal drilling and drilling for the storage of nuclear waste material, the drilling is within 100 metres of any controlled waters.	EIA is more likely to be required where the scale of the drilling operations involves development of a surface site of more than 5 hectares. Regard should be had to the likely wider impacts on surrounding hydrology and ecology. On its own, exploratory deep drilling is unlikely to require EIA. It would not be appropriate to require EIA for exploratory activity simply because it might eventually lead to some form of permanent activity.

Table (continued)

Column 1 Description of development	Column 2 Applicable thresholds and criteria	Column 3 Indicative thresholds and criteria
2. Extractive industry (continued)		
(e) Surface industrial installations for the extraction of coal, petroleum, natural gas and ores, as well as bituminous shale.	The area of the development exceeds 0.5 hectare.	The main considerations are likely to be the scale of development, emissions to air, discharges to water, the risk of accident and the arrangements for transporting the fuel. EIA is more likely to be required if the development is on a major scale (site of 10 hectares or more) or where production is expected to be substantial (e.g. more than 100 000 tonnes of petroleum per year).
3. Energy industry		
(a) Industrial installations for the production of electricity, steam and hot water (unless included in Schedule 1);	The area of the development exceeds 0.5 hectare.	EIA will normally be required for power stations which require approval from the Secretary of State at the Department of Trade and Industry (i.e. those with a thermal output of more than 50 megawatts). EIA is unlikely to be required for smaller new conventional power stations. Small stations using novel forms of generation should be considered carefully in line with the guidance in PPG 22 (Renewable Energy). The main considerations are likely to be the level of emissions to air, arrangements for the transport of fuel and any visual impact.
(b) Industrial installations for carrying gas, steam and hot water;	The area of the works exceeds 1 hectare.	
(c) Surface storage of natural gas;	(i) The area of any new building, deposit or structure exceeds 500 square metres; or (ii) a new building, deposit or structure is to be sited within 100 metres of any controlled waters.	In addition to the scale of the development, significant effects are likely to depend on discharges to water, emissions to air and risk of accidents. EIA is more likely to be required where it is proposed to store more than 100 000 tonnes of fuel. Smaller installations are unlikely to require EIA unless hazardous chemicals are stored.
(d) Underground storage of combustible gases;		
(e) Surface storage of fossil fuels;		
(f) Industrial briquetting of coal and lignite;	The area of new floorspace exceeds 1000 square metres.	As paragraph 4 – Production and processing of metals.

(g) Installations for the processing and storage of radioactive waste (unless included in Schedule 1);

(i) The area of new floorspace exceeds 1000 square metres; or

(ii) the installation resulting from the development will require an authorisation or the variation of an authorisation under the Radioactive Substances Act 1993.

EIA will normally be required for new installations whose primary purpose is to process and store radioactive waste, and which are located on sites not previously authorised for such use. In addition to the scale of any development, significant effects are likely to depend on the extent of routine discharges of radiation to the environment. In this context EIA is unlikely to be required for installations where the processing or storage of radioactive waste is incidental to the main purpose of the development (e.g. installations at hospitals or research facilities).

(h) Installations for hydroelectric energy production;

The installation is designed to produce more than 0.5 megawatts.

In addition to the physical scale of the development, particular regard should be had to the potential wider impacts on hydrology and ecology. EIA is more likely to be required for new hydroelectric developments which have more than 5 megawatts of generating capacity.

(i) Installations for the harnessing of wind power for energy production (wind farms).

(i) The development involves the installation of more than 2 turbines; or

(ii) the hub height of any turbine or height of any other structure exceeds 15 metres.

The likelihood of significant effects will generally depend on the scale of the development, and its visual impact, as well as potential noise impacts. EIA is more likely to be required for commercial developments of five or more turbines, or more than 5 megawatts of new generating capacity.

4. Production and processing of metals

(a) Installations for the production of pig iron or steel (primary or secondary fusion) including continuous casting;

The area of new floorspace exceeds 1000 square metres.

New manufacturing or industrial plants of the types listed in the Regulations, may well require EIA if the operational development covers a site of more than 10 hectares. Smaller developments are more likely to require EIA if they are expected to give rise to significant discharges of waste, emission of pollutants or operational noise. Among the factors to be taken into account in assessing the significance of such effects are:

• whether the development involves a process designated as a 'scheduled process' for the purpose of air pollution control;

(b) Installations for the processing of ferrous metals:

(i) hot-rolling mills;
(ii) smitheries with hammers;
(iii) application of protective fused metal coats;

Table (continued)

Column 1 Description of development	Column 2 Applicable thresholds and criteria	Column 3 Indicative thresholds and criteria
4. Production and processing of metals (continued)		
(c) Ferrous metal foundries;		• whether the process involves discharges to water which require the consent of the Environment Agency;
(d) Installations for the smelting, including the alloyage, of non-ferrous metals, excluding precious metals, including recovered products (refining, foundry casting, etc.);		• whether the installation would give rise to the presence of environmentally significant quantities of potentially hazardous or polluting substances;
(e) Installations for surface treatment of metals and plastic material using an electrolytic or chemical process;		• whether the process would give rise to radioactive or other hazardous waste;
(f) Manufacture and assembly of motor vehicles and manufacture of motor-vehicle engines;	The area of new floorspace exceeds 1000 square metres.	• whether the development would fall under Council Directive 96/82/EC on the control of major accident hazards involving dangerous substances (COMAH).
(g) Shipyards;		However, the need for a consent under other legislation is not itself a justification for EIA.
(h) Installations for the construction and repair of aircraft;		
(i) Manufacture of railway equipment;		
(j) Swaging by explosives;		
(k) Installations for the roasting and sintering of metallic ores.		

5. Mineral industry

(a) Coke ovens (dry coal distillation);

(b) Installations for the manufacture of cement;

(c) Installations for the production of asbestos and the manufacture of asbestos-based products (unless included in Schedule 1);

As for paragraph 4.

(d) Installations for the manufacture of glass including glass fibre;

(e) Installations for smelting mineral substances including the production of mineral fibres;

(f) Manufacture of ceramic products by burning, in particular roofing tiles, bricks, refractory bricks, tiles, stoneware or porcelain.

The area of new floorspace exceeds 1000 square metres.

6. Chemical industry (unless included in Schedule 1)

(a) Treatment of intermediate products and production of chemicals;

As for paragraph 4.

(b) Production of pesticides and pharmaceutical products, paint and varnishes, elastomers and peroxides;

The area of new floorspace exceeds 1000 square metres.

Table (continued)

Column 1 Description of development	Column 2 Applicable thresholds and criteria	Column 3 Indicative thresholds and criteria
6. Chemical industry (unless included in Schedule 1) (continued)		
(c) Storage facilities for petroleum, petrochemical and chemical products.	(i) The area of any new building or structure exceeds 0.05 hectare; or (ii) more than 200 tonnes of petroleum, petrochemical or chemical products are to be stored at any one time.	As for paragraph 4.
7. Food industry		
(a) Manufacture of vegetable and animal oils and fats;		As for paragraph 4.
(b) Packing and canning of animal and vegetable products;		
(c) Manufacture of dairy products;	The area of new floorspace exceeds 1000 square metres.	
(d) Brewing and malting;		
(e) Confectionery and syrup manufacture;		
(f) Installations for the slaughter of animals;		
(g) Industrial starch manufacturing installations;		
(h) Fish-meal and fish-oil factories;		
(i) Sugar factories.		

8. Textile, leather, wood and paper industries

(a) Industrial plants for the production of paper and board (unless included in Schedule 1);	
(b) Plants for the pre-treatment (operations such as washing, bleaching, mercerisation) or dyeing of fibres or textiles;	The area of new floorspace exceeds 1000 square metres.
(c) Plants for the tanning of hides and skins;	
(d) Cellulose-processing and production installations.	
	As for paragraph 4.

9. Rubber industry

Manufacture and treatment of elastomer-based products.	The area of new floorspace exceeds 1000 square metres.	As for paragraph 4.

10. Infrastructure projects

(a) Industrial estate development projects;	The area of the development exceeds 0.5 hectare.	EIA is more likely to be required if the site area of the new development is more than 20 hectares. In determining whether significant effects are likely, particular consideration should be given to the potential increase in traffic, emissions and noise.

Table (continued)

10. Infrastructure projects

Column 1 Description of development	Column 2 Applicable thresholds and criteria	Column 3 Indicative thresholds and criteria
(b) Urban development projects, including the construction of shopping centres and car parks, sports stadiums, leisure centres and multiplex cinemas;	The area of the development exceeds 0.5 hectare.	In addition to the physical scale of such developments, particular consideration should be given to the potential increase in traffic, emissions and noise. EIA is unlikely to be required for the redevelopment of land unless the new development is on a significantly greater scale than the previous use, or the types of impact are of a markedly different nature or there is a high level of contamination. Development proposed for sites which have not previously been intensively developed are more likely to require EIA if: • the site area of the scheme is more than 5 hectares; or • it would provide a total of more than 10 000 square metres of new commercial floorspace; or • the development would have significant urbanising effects in a previously non-urbanised area (e.g. a new development of more than 1000 dwellings).
(c) Construction of intermodal transshipment facilities and of intermodal terminals (unless included in Schedule 1);	The area of the development exceeds 0.5 hectare.	In addition to the physical scale of the development, particular impacts for consideration are increased traffic, noise, emissions to air and water. Developments of more than 5 hectares are more likely to require EIA.
(d) Construction of railways (unless included in Schedule 1);	The area of the works exceeds 1 hectare.	For linear transport schemes, the likelihood of significant effects will generally depend on the estimated emissions, traffic, noise and vibration and degree of visual intrusion and impact on the surrounding ecology. EIA is more likely to be required for new development over 2 kilometres in length.

(e)	Construction of airfields (unless included in Schedule 1);	(i) The development involves an extension to a runway; or (ii) the area of the works exceeds 1 hectare.	The main impacts to be considered in judging significance are noise, traffic generation and emissions. New permanent airfields will normally require EIA, as will major works (such as new runways or terminals with a site area of more than 10 hectares) at existing airports. Smaller scale development at existing airports is unlikely to require EIA unless it would lead to significant increases in air or road traffic.
(f)	Construction of roads (unless included in Schedule 1);	The area of the works exceeds 1 hectare.	As for paragraph 10(d).
(g)	Construction of harbours and port installations including fishing harbours (unless included in Schedule 1);	The area of the works exceeds 1 hectare.	Primary impacts for consideration are those on hydrology, ecology, noise and increased traffic. EIA is more likely to be required if the development is on a major scale (e.g. would cover a site of more than 10 hectares). Smaller developments may also have significant effects where they include a quay or pier which would extend beyond the high water mark or would affect wider coastal processes.
(h)	Inland-waterway construction not included in Schedule 1, canalisation and flood-relief works;	The area of the works exceeds 1 hectare.	The likelihood of significant impacts is likely to depend primarily on the potential wider impacts on the surrounding hydrology and ecology. EIA is more likely to be required for development of over 2 kilometres of canal. The impact of flood relief works is especially dependent on the nature of the location and the potential effects on the surrounding ecology and hydrology. Schemes for which the area of the works would exceed 5 hectares or which are more than 2 kilometres in length would normally require EIA.
(i)	Dams and other installations designed to hold water or store it on a long-term basis (unless included in Schedule 1);	The area of the works exceeds 1 hectare.	In considering such developments, particular regard should be had to the potential wider impacts on the hydrology and ecology, as well as to the physical scale of the development. EIA is likely to be required for any major new dam (e.g. where the construction site exceeds 20 hectares).

Table (continued)

Column 1 Description of development	Column 2 Applicable thresholds and criteria	Column 3 Indicative thresholds and criteria
10. Infrastructure projects		
(j) Tramways, elevated and underground railways, suspended lines or similar lines of a particular type, used exclusively or mainly for passenger transport;	The area of the works exceeds 1 hectare.	As for paragraph 10(d).
(k) Oil and gas pipe-line installations (unless included in Schedule 1);	(i) The area of the works exceeds 1 hectare; or, (ii) in the case of a gas pipe-line, the installation has a design operating pressure exceeding 7 bar gauge.	For underground pipe-lines, the major impact to be considered will generally be the disruption to the surrounding ecosystems during construction, while for overground pipe-lines visual impact will be a key consideration. EIA is more likely to be required for any pipe-line over 5 kilometres long. EIA is unlikely to be required for pipe-lines laid underneath a road, or for those installed entirely by means of tunnelling.
(l) Installations of long-distance aqueducts;		
(m) Coastal work to combat erosion and maritime works capable of altering the coast through the construction, for example, of dykes, moles, jetties and other sea defence works, excluding the maintenance and reconstruction of such works;	All development.	The impact of such works will depend largely on the nature of the particular site and the likely wider impacts on natural coastal processes outside the site. EIA will be more likely where the area of the works would exceed 1 hectare.

(n) Groundwater abstraction and artificial groundwater recharge schemes not included in Schedule 1; (o) Works for the transfer of water resources between river basins not included in Schedule 1;	The area of the works exceeds 1 hectare.	Impacts likely to be significant are those on hydrology and ecology. Developments of this sort can have significant effects on environments some kilometres distant. This is particularly important for wetland and other sites where the habitat and species are particularly dependent on an aquatic environment. EIA is likely to be required for developments where the area of the works exceeds 1 hectare.
(p) Motorway service areas.	The area of the development exceeds 0.5 hectare.	Impacts likely to be significant are traffic, noise, air quality, ecology and visual impact. EIA is more likely to be required for new motorway service areas which are proposed for previously undeveloped sites and if the proposed development would cover an area of more than 5 hectares.

11. Other projects

(a) Permanent racing and test tracks for motorised vehicles;	The area of the development exceeds 1 hectare.	Particular consideration should be given to the size, noise impacts, emissions and the potential traffic generation. EIA is more likely to be required for developments with a site area of 20 hectares or more.
(b) Installations for the disposal of waste (unless included in Schedule 1);	(i) The disposal is by incineration; or (ii) the area of the development exceeds 0.5 hectare; or (iii) the installation is to be sited within 100 metres of any controlled waters.	The likelihood of significant effects will generally depend on the scale of the development and the nature of the potential impact in terms of discharge, emissions or odour. For installations (including landfill sites) for the deposit, recovery and/or disposal of household, industrial and/or commercial wastes (as defined by the Controlled Waste Regulations 1992) EIA is more likely to be required where new capacity is created to hold more than 50 000 tonnes per year, or to hold waste on a site of 10 hectares or more. Sites taking smaller quantities of these wastes, sites seeking only to accept inert wastes (demolition rubble, etc.) or civic amenity sites, are unlikely to require EIA.

Table (continued)

Column 1 Description of development	Column 2 Applicable thresholds and criteria	Column 3 Indicative thresholds and criteria
11. Other projects (continued)		
(c) Waste water treatment plants (unless included in Schedule 1);	The area of the development exceeds 1000 square metres.	Particular consideration should be given to the size, treatment process, pollution and nuisance potential, topography, proximity of dwellings and the potential impact of traffic movements. EIA is more likely to be required if the development would be on a substantial scale (e.g. site area of more than 10 hectares) or if it would lead to significant discharges (e.g. capacity exceeding 100 000 population equivalent). EIA should not be required simply because a plant is on a scale which requires compliance with the Urban Waste Water Treatment Directive (91/271/EEC).
(d) Sludge-deposition sites;	(i) The area of deposit or storage exceeds 0.5 hectare; or (ii) a deposit is to be made or scrap stored within 100 metres of any controlled waters.	Similar considerations will apply for sewage sludge lagoons as for waste disposal installations. EIA is more likely to be required where the site is intended to hold more than 5000 cubic metres of sewage sludge.
(e) Storage of scrap iron, including scrap vehicles;		Major impacts from storage of scrap iron are likely to be discharges to soil, site noise and traffic generation. EIA is more likely to be required where it is proposed to store scrap on an area of 10 hectares or more.
(f) Test benches for engines, turbines or reactors;		As for paragraph 4.
(g) Installations for the manufacture of artificial mineral fibres;	The area of new floorspace exceeds 1000 square metres.	
(h) Installations for the recovery or destruction of explosive substances;		
(i) Knackers' yards.		

12. Tourism and leisure

(a) Ski-runs, ski-lifts and cable-cars and associated developments;	(i) The area of the works exceeds 1 hectare; or (ii) the height of any building or other structure exceeds 15 metres.	EIA is more likely to be required if the development is over 500 metres in length or if it requires a site of more than 5 hectares. In addition to any visual or ecological impacts, particular regard should also be had to the potential traffic generation.
(b) Marinas;	The area of the enclosed water surface exceeds 1000 square metres.	In assessing whether significant effects are likely, particular regard should be had to any wider impacts on natural coastal processes outside the site, as well as the potential noise and traffic generation. EIA is more likely to be required for large new marinas, for example where the proposal is for more than 300 berths (seawater site) or 100 berths (freshwater site). EIA is unlikely to be required where the development is located solely within an existing dock or basin.
(c) Holiday villages and hotel complexes outside urban areas and associated developments;	The area of the development exceeds 0.5 hectare.	In assessing the significance of tourism development, visual impacts, impacts on ecosystems and traffic generation will be key considerations. The effects of new theme parks are more likely to be significant if it is expected that they will generate more than 250 000 visitors per year.
(d) Theme parks;		EIA is likely to be required for major new tourism and leisure developments which require a site of more than 10 hectares. In particular, EIA is more likely to be required for holiday villages or hotel complexes with more than 300 bed spaces, or for permanent camp sites or caravan sites with more than 200 pitches.
(e) Permanent camp sites and caravan sites;	The area of the development exceeds 1 hectare.	
(f) Golf courses and associated developments.	The area of the development exceeds 1 hectare.	New 18-hole golf courses are likely to require EIA. The main impacts are likely to be those on the surrounding hydrology, ecosystems and landscape, as well as those from traffic generation. Developments at existing golf courses are unlikely to require EIA.

Table (continued)

Column 1 Description of development	Column 2 Applicable thresholds and criteria	Column 3 Indicative thresholds and criteria
13.		
(a) Any change to or extension of development of a description listed in Schedule 1 or in paragraphs 1 to 12 of Column 1 of this table, where that development is already authorised, executed or in the process of being executed, and the change or extension may have significant adverse effects on the environment;	(i) In relation to development of a description mentioned in Column 1 of this table, the thresholds and criteria in the corresponding part of Column 2 of this table applied to the change or extension (and not to the development as changed or extended). (ii) In relation to development of a description mentioned in a paragraph in Schedule 1 (see Appendix 1) indicated below, the thresholds and criteria in Column 2 of the paragraph of this table indicated below applied to the change or extension (and not to the development as changed or extended):	Development which comprises a change or extension requires EIA only if the change or extension is likely to have significant environmental effects. This should be considered in the light of the general guidance in DETR Circular 2/99 (Welsh Office Circular 11/99) and the indicative thresholds shown above in column 3. However, the significance of any effects must be considered in the context of the existing development. In some cases, repeated small extensions may be made to development. Quantified thresholds cannot easily deal with this kind of 'incremental' development. In such instances, it should be borne in mind that the column 3 thresholds are indicative only. An expansion of the same size as a previous expansion will not automatically lead to the same determination on the need for EIA because the environment may have altered since the question was last addressed.

Paragraph in Schedule 1	Paragraph of this table
1	6(a)
2(a)	3(a)
2(b)	3(g)
3	3(g)
4	4
5	5
6	6(a)
7(a)	10(d) (in relation to railways) or 10(e) (in relation to airports)
7(b) and (c)	10(f)
8(a)	10(h)
8(b)	10(g)
9	11(b)
10	11(b)
11	10(n)
12	10(o)
13	11(c)
14	2(e)
15	10(i)
16	10(k)
17	1(c)
18	8(a)
19	2(a)
20	6(c).
(b) Development of a description mentioned in Schedule 1 undertaken exclusively or mainly for the development and testing of new methods or products and not used for more than two years.	All development.

APPENDIX 4

Requirements of the Regulations as to the content of environmental statements

Below are the statutory provisions with respect to the content of environmental statements, as set out in Parts I and II of Schedule 4 to the Town and Country Planning (Environmental Impact Assessment) (England and Wales) Regulations 1999.

Under the definition in Regulation 2(1), 'environmental statement' means a statement:

(a) that includes such of the information referred to in Part I of Schedule 4 as is reasonably required to assess the environmental effects of the development and which the applicant can, having regard in particular to current knowledge and methods of assessment, reasonably be required to compile, but

(b) that includes at least the information referred to in Part II of Schedule 4.

Part I

1. Description of the development, including in particular:

(a) a description of the physical characteristics of the whole development and the land-use requirements during the construction and operational phases;

(b) a description of the main characteristics of the production processes, for instance, nature and quantity of the materials used;

(c) an estimate, by type and quantity, of expected residues and emissions (water, air and soil pollution, noise, vibration, light, heat, radiation, etc.) resulting from the operation of the proposed development.

2. An outline of the main alternatives studied by the applicant or appellant and an indication of the main reasons for his choice, taking into account the environmental effects.

3. A description of the aspects of the environment likely to be significantly affected by the development, including, in particular, population, fauna, flora, soil, water, air, climatic factors, material assets, including the architectural and archaeological heritage, landscape and the inter-relationship between the above factors.

4. A description of the likely significant effects of the development on the environment, which should cover the direct effects and any indirect, secondary, cumulative, short, medium and long-term, permanent and temporary, positive and negative effects of the development, resulting from:

   (a) the existence of the development;
   (b) the use of natural resources;
   (c) the emission of pollutants, the creation of nuisances and the elimination of waste,

   and the description by the applicant of the forecasting methods used to assess the effects on the environment.

5. A description of the measures envisaged to prevent, reduce and where possible offset any significant adverse effects on the environment.

6. A non-technical summary of the information provided under paragraphs 1 to 5 of this Part.

7. An indication of any difficulties (technical deficiencies or lack of know-how) encountered by the applicant in compiling the required information.

Part II

1. A description of the development comprising information on the site, design and size of the development.

2. A description of the measures envisaged in order to avoid, reduce and, if possible, remedy significant adverse effects.

3. The data required to identify and assess the main effects which the development is likely to have on the environment.

4. An outline of the main alternatives studied by the applicant or appellant and an indication of the main reasons for his choice, taking into account the environmental effects.

5. A non-technical summary of the information provided under paragraphs 1 to 4 of this Part.

APPENDIX 5
Checklist of matters to be considered for inclusion in an environmental statement

This checklist is intended as a guide to the subjects that need to be considered in the course of preparing an environmental statement. It is unlikely that all the items will be relevant to any one project. (See paragraphs 31 and 32 of the main text.)

The environmental effects of a development during its construction and commissioning phases should be considered separately from the effects arising whilst it is operational. Where the operational life of a development is expected to be limited, the effects of decommissioning or reinstating the land should also be considered separately.

Section 1

Information describing the project

1.1 Purpose and physical characteristics of the project, including details of proposed access and transport arrangements, and of numbers to be employed and where they will come from.

1.2 Land use requirements and other physical features of the project:

(a) during construction;
(b) when operational;
(c) after use has ceased (where appropriate).

1.3 Production processes and operational features of the project:

(a) type and quantities of raw materials, energy and other resources consumed;

(b) residues and emissions by type, quantity, composition and strength including:

(i) discharges to water;
(ii) emissions to air;
(iii) noise;
(iv) vibration;
(v) light;
(vi) heat;
(vii) radiation;
(viii) deposits/residues to land and soil;
(ix) others.

1.4 Main alternative sites and processes considered, where appropriate, and reasons for final choice.

Section 2

Information describing the site and its environment

Physical features

2.1 Population — proximity and numbers.

2.2 Flora and fauna (including both habitats and species) — in particular, protected species and their habitats.

2.3 Soil: agricultural quality, geology and geomorphology.

2.4 Water: aquifers, water courses, shoreline, including the type, quantity, composition and strength of any existing discharges.

2.5 Air: climatic factors, air quality, etc.

2.6 Architectural and historic heritage, archaeological sites and features, and other material assets.

2.7 Landscape and topography.

2.8 Recreational uses.

2.9 Any other relevant environmental features.

The policy framework

2.10 Where applicable, the information considered under this section should include all relevant statutory designations such as national nature reserves, sites of special scientific interest, national parks, areas of outstanding natural beauty, heritage coasts, regional parks, country parks and designated green belt, local nature reserves, areas affected by tree preservation orders, water protection zones, conservation areas, listed buildings, scheduled ancient monuments, and designated areas of archaeological importance. It should also include references to relevant national policies (including Planning Policy Guidance notes and Planning Policy Statements) and to regional and local plans and policies (including approved or emerging development plans).

2.11 Reference should also be made to international designations, e.g. those under the EC 'Wild Birds' or 'Habitats' Directives, the Biodiversity Convention and the Ramsar Convention.

Section 3

Assessment of effects

Including direct and indirect, secondary, cumulative, short, medium and long-term, permanent and temporary, positive and negative effects of the project.

Effects on human beings, buildings and man-made features

3.1 Change in population arising from the development, and consequential environment effects.

3.2 Visual effects of the development on the surrounding area and landscape.

3.3 Levels and effects of emissions from the development during normal operation.

3.4 Levels and effects of noise from the development.

3.5 Effects of the development on local roads and transport.

3.6 Effects of the development on buildings, the architectural and historic heritage, archaeological features, and other human artefacts, e.g. through pollutants, visual intrusion, vibration.

Effects on flora, fauna and geology
3.7 Loss of, and damage to, habitats and plant and animal species.

3.8 Loss of, and damage to, geological, palaeontological and physiographic features.

3.9 Other ecological consequences.

Effects on land
3.10 Physical effects of the development, e.g. change in local topography, effect of earth-moving on stability, soil erosion, etc.

3.11 Effects of chemical emissions and deposits on soil of site and surrounding land.

3.12 Land use/resource effects:

(a) quality and quantity of agricultural land to be taken;
(b) sterilisation of mineral resources;
(c) other alternative uses of the site, including the 'do nothing' option;
(d) effect on surrounding land uses including agriculture;
(e) waste disposal.

Effects on water
3.13 Effects of development on drainage pattern in the area.

3.14 Changes to other hydrographic characteristics, e.g. groundwater level, water courses, flow of underground water.

3.15 Effects on coastal or estuarine hydrology.

3.16 Effects of pollutants, waste, etc. on water quality.

Effects on air and climate
3.17 Level and concentration of chemical emissions and their environmental effects.

3.18 Particulate matter.

3.19 Offensive odours.

3.20 Any other climatic effects.

Other indirect and secondary effects associated with the project
3.21 Effects from traffic (road, rail, air, water) related to the development.

3.22 Effects arising from the extraction and consumption of materials, water, energy or other resources by the development.

3.23 Effects of other development associated with the project, e.g. new roads, sewers, housing, power lines, pipe-lines, telecommunications, etc.

3.24 Effects of association of the development with other existing or proposed development.

3.25 Secondary effects resulting from the interaction of separate direct effects listed above.

Section 4

Mitigating measures

4.1 Where significant adverse effects are identified, a description of the measures to be taken to avoid, reduce or remedy those effects, e.g:

(a) site planning;
(b) technical measures, e.g:

    (i) process selection;
    (ii) recycling;
    (iii) pollution control and treatment;
    (iv) containment (e.g, bunding of storage vessels).

(c) aesthetic and ecological measures, e.g:

    (i) mounding;
    (ii) design, colour, etc;
    (iii) landscaping;

(iv)   tree plantings;
(v)   measures to preserve particular habitats or create alternative habitats;
(vi)   recording of archaeological sites;
(vii)   measures to safeguard historic buildings or sites.

4.2   Assessment of the likely effectiveness of mitigating measures.

Section 5

Risk of accidents and hazardous development

5.1   Risk of accidents as such is not covered in the EIA Directive or, consequently, in the implementing Regulations. However, when the proposed development involves materials that could be harmful to the environment (including people) in the event of an accident, the environmental statement should include an indication of the preventive measures that will be adopted so that such an occurrence is not likely to have a significant effect. This could, where appropriate, include reference to compliance with Health and Safety legislation.

5.2   There are separate arrangements in force relating to the keeping or use of hazardous substances and the Health and Safety Executive provides local planning authorities with expert advice about risk assessment on any planning application involving a hazardous installation.

5.3   Nevertheless, it is desirable that, wherever possible, the risk of accidents and the general environmental effects of developments should be considered together, and developers and planning authorities should bear this in mind.

APPENDIX 6
List of statutory consultees where environmental impact assessment is carried out

See the definition of 'the consultation bodies' in Regulation 2(1) of the Town and Country Planning (Environmental Impact Assessment) (England and Wales) Regulations 1999.

(1) Any principal council for the area where the land is situated, if not the relevant planning authority.

(2) The Countryside Agency and English Nature (where the land is situated in England).

(3) The Countryside Council for Wales (where the land is situated in Wales).

(4) The Environment Agency.

(5) The Secretary of State for Wales (where the land is situated in Wales).

(6) The bodies which would be statutory consultees under Article 10 of the Town and Country Planning (General Development Procedure) Order 1995 (GDPO) for any planning application for the proposed development.

For ease of reference, Article 10 of the GDPO is reproduced below:

Consultations before the grant of permission

10. (1) Before granting planning permission for development which, in their opinion, falls within a category set out in the table below, a local planning authority shall consult the authority or person mentioned in relation to that category, except where:

(i) the local planning authority are the authority so mentioned;

(ii) the local planning authority are required to consult the authority so mentioned under articles 11 or 12; or

(iii) the authority or person so mentioned has advised the local planning authority that they do not wish to be consulted.

[(1A) The exception in article 10(1)(iii) shall not apply where, in the opinion of the local planning authority, development falls within paragraph (zb) of the table below.]

Table

Para	Description of development	Consultee
(a)	Development likely to affect land in Greater London or in a metropolitan county [or, in relation to Wales, land in the area of another local planning authority]	The local planning authority concerned
(b)	Development likely to affect land in a non-metropolitan county [in England], other than land in a National Park	The district planning authority concerned
(c)	Development likely to affect land in a National Park [in England]	The county planning authority concerned
(d)	Development within an area which has been notified to the local planning authority by the Health and Safety Executive for the purpose of this provision because of the presence within the vicinity of toxic, highly reactive, explosive or inflammable substances and which involves the provision of:	The Health and Safety Executive

(i) residential accommodation;
(ii) more than 250 square metres of retail floor space;
(iii) more than 500 square metres of office floor space; or
(iv) more than 750 square metres of floor space to be used for an industrial process,

or which is otherwise likely to result in a material increase in the number of persons working within or visiting the notified area

Table (continued)

Para	Description of development	Consultee
(e)	Development likely to result in a material increase in the volume or a material change in the character of traffic: (i) entering or leaving a trunk road; or (ii) using a level crossing over a railway	In England, the Secretary of State for Transport and, in Wales, the Secretary of State for Wales The operator of the network which includes or consists of the railway in question, and in England, the Secretary of State for Transport and, in Wales, the Secretary of State for Wales
(f)	Development likely to result in a material increase in the volume or a material change in the character of traffic entering or leaving a classified road or proposed highway	The local highway authority concerned
(g)	Development likely to prejudice the improvement or construction of a classified road or proposed highway	The local highway authority concerned
(h)	Development involving: (i) the formation, laying out or alteration of any means of access to a highway (other than a trunk road); or (ii) the construction of a highway or private means of access to premises affording access to a road in relation to which a toll order is in force	The local highway authority concerned The local highway authority concerned, and in the case of a road subject to a concession, the concessionaire
(i)	Development which consists of or includes the laying out or construction of a new street	The local highway authority
(j)	Development which involves the provision of a building or pipe-line in an area of coal working notified by the Coal Authority to the local planning authority	The Coal Authority
(k)	Development involving or including mining operations	The Environment Agency
(l)	Development within three kilometres of Windsor Castle, Windsor Great Park, or Windsor Home Park, or within 800 metres of any other royal palace or park, which might affect the amenities (including security) of that palace or park	The Secretary of State for National Heritage
(m)	Development of land in Greater London involving the demolition, in whole or part, or the material alteration of a listed building	The Historic Buildings and Monuments Commission for England

Table (continued)

Para	Description of development	Consultee
(n)	Development likely to affect the site of a scheduled monument	In England, the Historic Buildings and Monuments Commission for England, and, in Wales, the Secretary of State for Wales
(o)	Development likely to affect any garden or park of special historic interest which is registered in accordance with section 8C of the Historic Buildings and Ancient Monuments Act 1953 (register of gardens) and which is classified as Grade I or Grade II*	The Historic Buildings and Monuments Commission for England
(p)	Development involving the carrying out of works or operations in the bed of or on the banks of a river or stream	The Environment Agency
(q)	Development for the purpose of refining or storing mineral oils and other derivatives	The Environment Agency
(r)	Development involving the use of land for the deposit of refuse or waste	The Environment Agency
(s)	Development relating to the retention, treatment or disposal of sewage, trade waste, slurry or sludge (other than the laying of sewers, the construction of pumphouses in a line of sewers, the construction of septic tanks and cesspools serving single dwellinghouses or single caravans or single buildings in which not more than ten people will normally reside, work or congregate, and works ancillary thereto)	The Environment Agency
(t)	Development relating to the use of land as a cemetery	The Environment Agency
(u)	Development: (i) in or likely to affect a site of special scientific interest of which notification has been given, or has effect as if given, to the local planning authority by the Nature Conservancy Council for England or the Countryside Council for Wales, in accordance with section 28 of the Wildlife and Countryside Act 1981 (areas of special scientific interest); or (ii) within an area which has been notified to the local planning authority by the Nature Conservancy Council for England or the Countryside Council for Wales, and which is within two kilometres of a site of special scientific interest of which notification has been given or has effect as if given as aforesaid	The Council which gave, or is to be regarded as having given, the notice

Table (continued)

Para	Description of development	Consultee
(v)	Development involving any land on which there is a theatre	The Theatres Trust
(w)	Development which is not for agricultural purposes and is not in accordance with the provisions of a development plan and involves: (i) the loss of not less than 20 hectares of grades 1, 2 or 3a agricultural land which is for the time being used (or was last used) for agricultural purposes; or (ii) the loss of less than 20 hectares of grades 1, 2 or 3a agricultural land which is for the time being used (or was last used) for agricultural purposes, in circumstances in which the development is likely to lead to a further loss of agricultural land amounting cumulatively to 20 hectares or more	In England, the Minister of Agriculture, Fisheries and Food and, in Wales, the Secretary of State for Wales
(x)	Development within 250 metres of land which: (i) is or has, at any time in the 30 years before the relevant application, been used for the deposit of refuse or waste; and (ii) has been notified to the local planning authority by the waste regulation authority for the purposes of this provision	The waste regulation authority concerned
(y)	Development for the purposes of fish farming	The Environment Agency
(z)	Development which: (i) is likely to prejudice the use, or lead to the loss of use, of land being used as a playing field; or (ii) is on land which has been: (aa) used as a playing field at any time in the five years before the making of the relevant application and which remains undeveloped; or (bb) allocated for use as a playing field in a development plan or in proposals for such a plan or its alteration or replacement; or (iii) involves the replacement of the grass surface of a playing pitch on a playing field with an artificial, man-made or composite surface	In England, the Sports Council for England; in Wales, the Sports Council for Wales

Table (continued)

Para	Description of development	Consultee
(za) [1]Development likely to affect: (i) any inland waterway (whether natural or artificial) or reservoir owned or managed by the British Waterways Board; or (ii) any canal feeder channel, watercourse, let off or culvert, which is within an area which has been notified for the purposes of this provision to the local planning authority by the British Waterways Board	The British Waterways Board
(zb) Development: (i) involving the siting of new establishments; or (ii) consisting of modifications to existing establishments which could have significant repercussions on major-accident hazards; or (iii) including transport links, locations frequented by the public and residential areas in the vicinity of existing establishments, where the siting or development is such as to increase the risk or consequences of a major accident	The Health and Safety Executive and the Environment Agency, and, where it appears to the local planning authority that an area of particular natural sensitivity or interest may be affected, in England, the Nature Conservancy Council for England, or in Wales, the Countryside Council for Wales.

[1] Inserted by SI 1997 No. 858. The amendment does not affect any planning application made before 1 July 1997.

(2) In the above table:

(a) in paragraph (d)(iv), 'industrial process' means a process for or incidental to any of the following purposes:

(i) the making of any article or part of any article (including a ship or vessel, or a film, video or sound recording);

(ii) the altering, repairing, maintaining, ornamenting, finishing, cleaning, washing, packing, canning, adapting for sale, breaking up or demolition of any article; or

(iii) the getting, dressing or treatment of minerals in the course of any trade or business other than agriculture, and other than a process carried out on land used as a mine or adjacent to and occupied together with a mine (and, in this sub-paragraph, 'mine' means any site on which mining operations are carried out);

(b) in paragraph (e)(ii), 'network' and 'operator' have the same meaning as in Part I of the Railways Act 1993 (the provision of railway services);

(c) in paragraphs (f) and (g), 'classified road' means a highway or proposed highway which:

  (i) is a classified road or a principal road by virtue of section 12(1) of the Highways Act 1980 (general provision as to principal and classified roads); or

  (ii) is classified for the purposes of any enactment by the Secretary of State by virtue of section 12(3) of that Act;

(d) in paragraph (h), 'concessionaire', 'road subject to a concession' and 'toll order' have the same meaning as in Part I of the New Roads and Street Works Act 1991 (new roads in England and Wales);

(e) in paragraph (i), 'street' has the same meaning as in section 48(1) of the New Roads and Street Works Act 1991 (streets, street works and undertakers), and 'new street' includes a continuation of an existing street;

(f) in paragraph (m), 'listed building' has the same meaning as in section 1 of the Planning (Listed Buildings and Conservation Areas) Act 1990 (listing of buildings of special architectural or historic interest);

(g) in paragraph (n), 'scheduled monument' has the same meaning as in section 1(11) of the Ancient Monuments and Archaeological Areas Act 1979 (schedule of monuments);

(h) in paragraph (s), 'slurry' means animal faeces and urine (whether or not water has been added for handling), and 'caravan' has the same meaning as for the purposes of Part I of the Caravan Sites and Control of Development Act 1960 (caravan sites);

(i) in paragraph (u), 'site of special scientific interest' means land to which section 28(1) of the Wildlife and Countryside Act 1981 (areas of special scientific interest) applies;

(j) in paragraph (v), 'theatre' has the same meaning as in section 5 of the Theatres Trust Act 1976 (interpretation);

(k) in paragraph (x), 'waste regulation authority' has the same meaning as in section 30(1) of the Environmental Protection Act 1990 (authorities for purposes of Part II);

[(l) in paragraph (z):

   (i) 'playing field' means the whole of a site which encompasses at least one playing pitch;

   (ii) 'playing pitch' means a delineated area which, together with any run-off area, is of 0.4 hectares or more, and which is used for association football, American football, rugby, cricket, hockey, lacrosse, rounders, baseball, softball, Australian football, Gaelic football, shinty, hurling, polo or cycle polo] [; and

(m) the expressions used in paragraph (zb), have the same meaning as in Council Directive 96/82/EC on the control of major accident hazards involving dangerous substances.]

(3) The Secretary of State may give directions to a local planning authority requiring that authority to consult any person or body named in the directions, in any case or class of case specified in the directions.

(4) Where, by or under this article, a local planning authority are required to consult any person or body ('the consultee') before granting planning permission:

   (a) they shall, unless an applicant has served a copy of an application for planning permission on the consultee, give notice of the application to the consultee; and

   (b) they shall not determine the application until at least 14 days after the date on which notice is given under paragraph (a) or, if earlier, 14 days after the date of service of a copy of the application on the consultee by the applicant.

(5) The local planning authority shall, in determining the application, take into account any representations received from a consultee.

APPENDIX 7
Flow charts illustrating the main procedural stages

Note: references in this appendix to the Secretary of State mean, in Wales, the National Assembly.

1. PRE-APPLICATION REQUEST BY DEVELOPER TO LOCAL PLANNING AUTHORITY FOR SCREENING OPINION

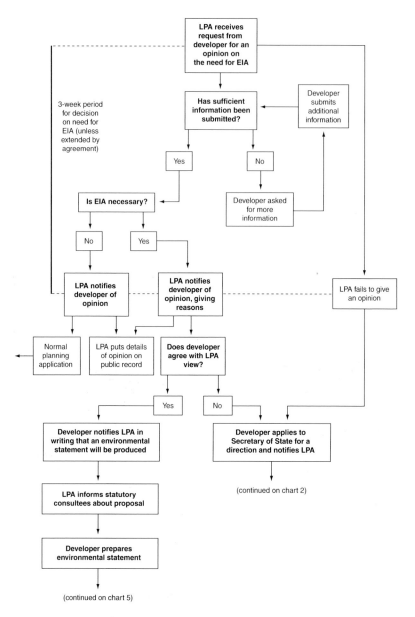

LPA receives request from developer for an opinion on the need for EIA

3-week period for decision on need for EIA (unless extended by agreement)

Has sufficient information been submitted?

Developer submits additional information

Yes

No

Is EIA necessary?

Developer asked for more information

No

Yes

LPA notifies developer of opinion

LPA notifies developer of opinion, giving reasons

LPA fails to give an opinion

Normal planning application

LPA puts details of opinion on public record

Does developer agree with LPA view?

Yes

No

Developer notifies LPA in writing that an environmental statement will be produced

Developer applies to Secretary of State for a direction and notifies LPA

LPA informs statutory consultees about proposal

(continued on chart 2)

Developer prepares environmental statement

(continued on chart 5)

106

2. PRE-APPLICATION REQUEST TO SECRETARY OF STATE FOR SCREENING DIRECTION

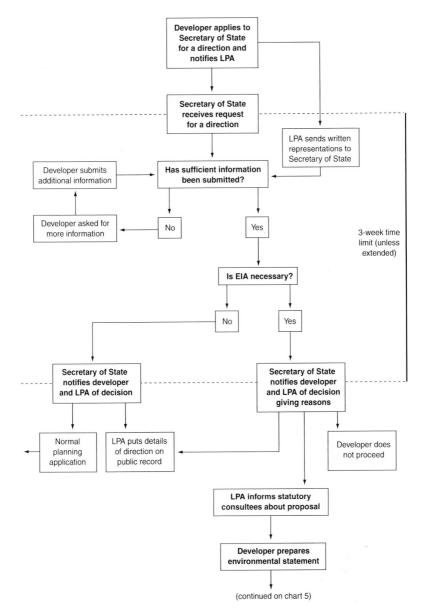

Developer applies to Secretary of State for a direction and notifies LPA

Secretary of State receives request for a direction

LPA sends written representations to Secretary of State

Developer submits additional information

Has sufficient information been submitted?

Developer asked for more information

No

Yes

3-week time limit (unless extended)

Is EIA necessary?

No

Yes

Secretary of State notifies developer and LPA of decision

Secretary of State notifies developer and LPA of decision giving reasons

Normal planning application

LPA puts details of direction on public record

Developer does not proceed

LPA informs statutory consultees about proposal

Developer prepares environmental statement

(continued on chart 5)

3. APPLICATION BY DEVELOPER TO LOCAL PLANNING AUTHORITY FOR SCOPING OPINION

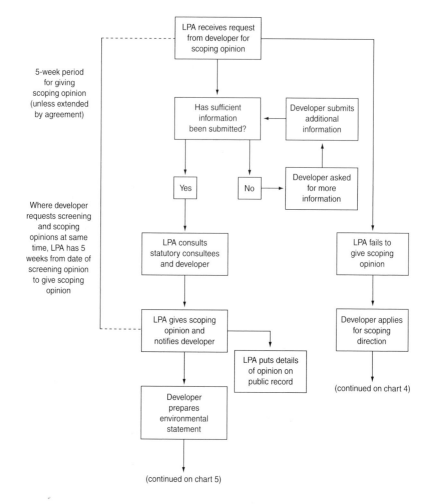

5-week period
for giving
scoping opinion
(unless extended
by agreement)

LPA receives request from developer for scoping opinion

Has sufficient information been submitted?

Developer submits additional information

Yes

No

Developer asked for more information

Where developer requests screening and scoping opinions at same time, LPA has 5 weeks from date of screening opinion to give scoping opinion

LPA consults statutory consultees and developer

LPA fails to give scoping opinion

LPA gives scoping opinion and notifies developer

LPA puts details of opinion on public record

Developer applies for scoping direction

Developer prepares environmental statement

(continued on chart 4)

(continued on chart 5)

4. APPLICATION TO SECRETARY OF STATE FOR SCOPING DIRECTION

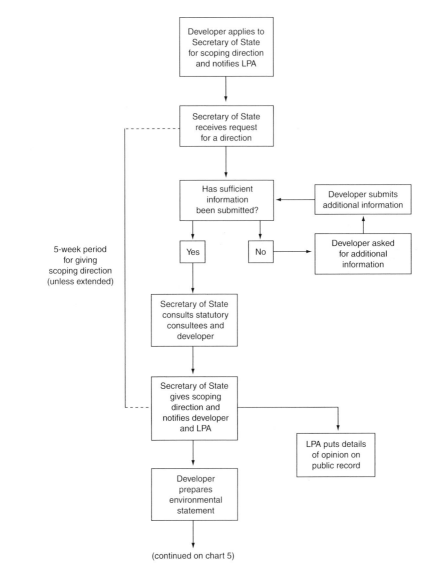

Developer applies to
Secretary of State
for scoping direction
and notifies LPA

Secretary of State
receives request
for a direction

Has sufficient
information
been submitted?

Developer submits
additional information

5-week period
for giving
scoping direction
(unless extended)

Yes

No

Developer asked
for additional
information

Secretary of State
consults statutory
consultees and
developer

Secretary of State
gives scoping
direction and
notifies developer
and LPA

LPA puts details
of opinion on
public record

Developer
prepares
environmental
statement

(continued on chart 5)

5. SUBMISSION OF ENVIRONMENTAL STATEMENT TO LOCAL PLANNING AUTHORITY IN CONJUNCTION WITH PLANNING APPLICATION

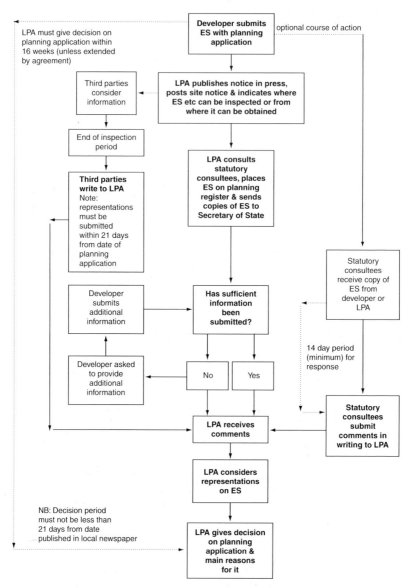

APPENDIX 8
UK Statutory Instruments and other official publications on environmental impact assessment

1. Council Directive 97/11/EC of 3 March 1997 is printed in the Official Journal of the European Communities, page no. L73/5 dated 14.3.97. It amends Directive 85/337/EEC on the assessment of certain public and private projects on the environment, which is printed in the Official Journal on page no. L175/40 dated 5.7.85. A consolidated version of the Directive as amended is reproduced in Appendix 1 to this booklet.

2. The following Regulations implementing Directive 85/337/EEC as amended by Directive 97/11/EC have been made:

    (i)   Town and Country Planning (Environmental Impact Assessment) (England and Wales) Regulations 1999 (SI No. 293)

    (ii)   Environmental Impact Assessment (Scotland) Regulations 1999 (SSI 1)

    (iii) Planning (Environmental Impact Assessment) Regulations (Northern Ireland) 1999 (SR No. 73)

    (iv)   Environmental Impact Assessment (Forestry) (England and Wales) Regulations 1999 (SI No. 2228)

    (v)   Environmental Impact Assessment (Forestry) (Scotland) Regulations 1999 (SSI 43)

    (vi)   Environmental Impact Assessment (Forestry) Regulations (Northern Ireland) 2000 (SR No. 84)

    (vii) Environmental Impact Assessment (Land Drainage Improvement Works) Regulations 1999 (SI No. 1783)

(viii)   Environmental Impact Assessment (Fish Farming in Marine Waters) Regulations 1999 (SI No. 367)

(ix)   Environmental Impact Assessment (Fish Farming in Marine Waters) Regulations (Northern Ireland) 1999 (SR No. 415)

(x)   Highways (Assessment of Environmental Effects) Regulations 1999 (SI No. 369)

(xi)   Roads (Environmental Impact Assessment) Regulations (Northern Ireland) 1999 (SR 89)

(xii)   Harbour Works (Environmental Impact Assessment) Regulations 1999 (SI No. 3445)

(xiii)   Nuclear Reactors (Environmental Impact Assessment for Decommissioning) Regulations 1999 (SI No. 2892)

(xiv)   Offshore Petroleum Production and Pipe-lines (Assessment of Environmental Effects) Regulations 1999 (SI No. 360)

(xv)   Public Gas Transporter Pipe-line Works (Environmental Impact Assessment) Regulations 1999 (SI No. 1672)

(xvi)   Electricity Works (Assessment of Environmental Effects) Regulations 2000 (SI No. 1927)

(xvii)   Pipe-line Works (Environmental Impact Assessment) Regulations 2000 (SI No. 1928)

(xviii)   Transport and Works (Applications and Objections Procedure) (England and Wales) Rules 2000 (SI No. 2190)

(xix)   Town and Country Planning (Environmental Impact Assessment) (England and Wales) (Amendment) Regulations 2000 (SI No. 2867)

(xx)   Harbour Works (Environmental Impact Assessment) (Amendment) Regulations 2000 (SI No. 2391)

(xxi)   Electricity Works (Environmental Impact Assessment) (Scotland) Regulations 2000 (SSI No. 320)

(xxii)   Transport and Works (Assessment of Environmental Effects) Regulations 2000 (England and Wales) (SI No. 3199)

(xxiii)   Drainage (Environmental Impact Assessment) Regulations (Northern Ireland) 2001 (SR No. 394)

(xxiv)   The Electricity Act 1989 (Requirement of Consent for Offshore Wind and Water Driven Generating Stations) (England and Wales) Order 2001 (SI No. 3642)

(xxv)   The Environmental Impact Assessment (Uncultivated Land and Semi-natural Areas) (England) Regulations 2001 (SI No. 3966)

(xxvi) Environmental Impact Assessment (Uncultivated Land and Semi-Natural Areas) Regulations (Northern Ireland) 2001 (SR No. 435)

(xxvii) The Environmental Impact Assessment (Uncultivated Land and Semi-Natural Areas) (Scotland) Regulations 2002 (SSI No. 6)

(xxviii) Environmental Impact Assessment (Scotland) Amendment Regulations 2002 (SSI No. 324)

(xxix) The Electricity Act 1989 (Requirement of Consent for Offshore Wind Generating Stations) (Scotland) Order 2002 (SSI No. 407)

(xxx) Environmental Impact Assessment (Uncultivated Land and Semi-Natural Areas) Regulations (Wales) 2002 (WSI No. 2127)

(xxxi) Harbour Works (Environmental Impact Assessment) Regulations (Northern Ireland) 2003 (SR No. 136)

(xxxii) The Water Resources (Environmental Impact Assessment) Regulations 2003 (SI No. 164)

3. The following Regulations to implement Directive 85/337/EEC as amended by Directive 97/11/EC are in the course of preparation:

(i) Environmental Impact Assessment and Habitats (Extraction of Minerals by Marine Dredging) Regulations

(ii) The Water Resources (Environmental Impact Assessment) Regulations (Northern Ireland) 2004

Guidance

(i) DETR Circular 02/99

(ii) Welsh Office Circular 11/99

(iii) Scottish Executive Development Department Circular 15/1999

(iv) Scottish Executive Development Department Planning Advice Note 58
[(iii) and (iv) are available free of charge on tel: 0131 2444 7066 or email: planningdivision@scotland.gov.uk]

(v) Planning Service (Northern Ireland) Development Control Advice Note 10 (Revised 1999)

(vi) Preparation of Environmental Statements for Planning Projects that require Environmental Assessment (Good Practice Guide, HMSO 1995)

(vii)   Evaluation of Environmental Information for Planning Projects (Good Practice Guide, HMSO, 1994)

(viii)   Environmental Impact Assessment of Forestry Projects, issued by the Forestry Commission

(ix)   Guide to The Environmental Impact Assessment (Fish Farming in Marine Waters) Regulations 1999 (Scottish Executive Rural Affairs Department, March 1999)

(x)   Environmental Assessment Guidance Manual for Marine Salmon Farmers (Crown Estate Commissioners in consultation with Scottish Executive)

(xi)   A guide to TWA Procedures (available on www.planning.dtlr.gov.uk/twa92/index.htm).

(xii)   Guidance and advice on the Uncultivated land Regulations is available at www.defra.gov.uk/eia, tel. 0800 028 2140; or write to EIA Unit, Rural Development Service, Defra, Coley Park, Reading RG1 6DT, or email eia.england@defra.gsi.gov.uk

(xiii)   Guidance on the Water Resources Regulations is available at www.environment-agency.gov.uk/subjects/waterres/469477/?version=1&lang=_e

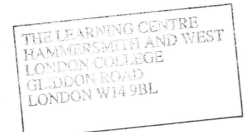